THE COSMIC CONNECTION

JEFF KANIPE

THE COSMIC CONNECTION

HOW
ASTRONOMICAL
EVENTS
IMPACT LIFE
ON EARTH

Prometheus Books

59 John Glenn Drive
Amherst, New York 14228–2119

Published 2009 by Prometheus Books

Inquiries should be addressed to
Prometheus Books
59 John Glenn Drive
Amherst, New York 14228–2119
VOICE: 716–691–0133, ext. 210
FAX: 716–691–0137
WWW.PROMETHEUSBOOKS.COM

13 12 11 10 09   5 4 3 2 1

Library of Congress Cataloging-in-Publication Data

Kanipe, Jeff, 1953–
    The cosmic connection : how astronomical events impact life on Earth / Jeff Kanipe.
        p. cm.
    Includes bibliographical references and index.
    ISBN 978–1–59102–667–9 (hardcover)
    1. Astronomy—Miscellanea. I. Title.

QB52.K26 2008
520—dc22

                                                                    2008031877

Printed in the United States of America on acid-free paper

To the memory of **Harlan J. Smith,**

director of McDonald Observatory,
1963–1989

CONTENTS

Introduction   9

1. Tilt-a-Whirl World   15

2. An Imperfect Sun   39

3. A Temperamental Sun   65

4. At Any Time   87

5. Keep Your Distance   125

6. There Goes the Neighborhood   153

7. Keep Watching the Skies!   185

8. Exotica   223

CONTENTS

9. The Ultimate Cosmic Connection 251

Acknowledgments 265

Notes 267

Index 289

INTRODUCTION

As sentient creatures, we humans have been looking at the heavens for tens of thousands of years, give or take, and yet civilization's reach has made little more than a chalk jot on the great wall of time. But during that time we've gone from fashioning stone tools to achieving scientific and technological breakthroughs bordering on the miraculous. In the twentieth century alone, old-timers could tell striplings that they lived in a time before the automobile and the airplane but also lived to see artificial satellites moving silently among the stars of the night sky, astronauts ambling on the Moon, and pictures taken from the surface of Mars. We now live in a period of on-the-verge science, typified by exciting new insights in quantum mechanics, genetics, nanotechnology, and stem cell research. One can only wonder what the next 10,000 years will bring!

Alas, if only humankind could be characterized by such positive strides. Unfortunately, there is no getting around the fact that our forward momentum has too often been severely compromised by our inherent impulses toward infamy beyond comprehension.

While I certainly don't agree with W. S. Gilbert that humankind is "Nature's sole mistake," it does seem, sometimes, as if we are Nature's experiment—one that, hopefully, will turn out to be a success.

Our ascendancy as a species is usually credited to Darwinian processes, such as passing along traits from one generation to the next, genetic mutations that improve an organism's chances of survival, successful adaptations of organisms to different regions or environments, and the flourishing of one species over another. Nevertheless, evolution alone is not enough to explain the ascension of the human race on this amazing planet. In its most sweeping terms, life also results from conditions not of our world but of our universe. Some of these, like the motion of the Sun through the Galaxy, might be said to be mechanical in nature. Others, like the impact of a comet or asteroid on Earth, could be considered random, capricious events. In any case, evolution would have taken a two-dimensional path without them.

Long before protohumans dropped from the trees and began playing with fire and scribbling on cave walls, even before the Sun emerged from its chrysalis of dust and irradiated its nascent brood of planets, numberless and nameless astronomical events steered the course and direction of the Earth and its emerging life-forms. These astronomical factors contributed to giving the Earth its early advantage in the cosmic game of life. Others can be credited with maintaining, even nurturing, that life. These include the Earth's location in space, its stable orbit around the Sun, the fact that it has a satellite that prevents the planet's rotational axis from tipping wildly, and the even temperament of its star. We can literally thank our lucky stars we are here because as any astronomer will tell you, the developmental history of any planet, much less one with life sticking to it, is fraught with peril and uncertainty. You don't have to look very far into the Galaxy, or even our own solar system, to realize that were the Earth located someplace else, even by a small distance, our so-called habitable zone might not be so habitable. Stars are not, by and large, life-form-friendly. The vast majority are too cool to nurture anything beyond the most rudimentary organisms, while a smaller fraction are too short-lived and volatile to host habitable planets. In any galaxy, one of the biggest factors for successfully engendering life on an earthlike planet (never mind self-aware life) is that it maintains a comfortable distance from its home star, and especially from other potentially deadly stars in the vicinity, whose explosive deaths could result in the destruction of the atmosphere or in undesirable mutagenetic effects. Civilization might not have gotten very far had the Star of Bethlehem been a nearby supernova.

To be sure, no astronomical event can be said to have had any adverse biological or ecological impact at all during the last 2,000 or even 10,000 years. Comets may have buzzed past, asteroids skittered off our atmosphere, and stars exploded near enough to light up our skies during the day—but no cosmic event has lifted a deadly hand toward Earth and probably won't in our lifetime or a hundred lifetimes. Only when you assume an astronomical perspective does it become obvious that there is no such thing as cosmic isolation. A civilization that lives long enough won't have to go to the stars; the stars will come to it.

The effects of these cosmic interventions both giveth and taketh away. Dying stars are the seeds of life. Nitrogen, found in complex organic molecules called polycyclic aromatic nitrogen heterocycles, or PANHs for the rest of us, forms in abundance around dying stars and star-forming regions. Since interstellar space is full of dying stars and their nebulae, PANHs must exist in every corner of the Galaxy, including our own.[1] And indeed they do, and not just in our Galaxy but throughout the universe. More important, however, since much of the chemistry of life requires organic molecules containing nitrogen, it stands to reason that the nitrogen in our DNA was sown across space by previous generations of indiscriminate, dying stars.

Comets, too, may have played a role in sowing life on Earth. When a comet called LINEAR strayed too close to the Sun in 2000 and broke up into a chain of icy debris, astronomers analyzed the fragments and discovered that they were not only made up of water but that the water had a chemical composition similar to water on Earth. This finding bolstered what has been a controversial theory, that comets in the early solar system seeded the Earth billions of years ago with water and essential molecular building blocks. These comets would not have formed in the frozen depths of the outer solar system but in the warmer environs around the planet Jupiter, where reactions in the primordial gases would have produced greater amounts of organic molecules as well as a just-right balance of "heavy water," which is composed of hydrogen, oxygen, and the form, or isotope, of hydrogen called deuterium.[2] On the other hand, it's also apparent that the stars fell on the dinosaurs sixty-five million years ago in the form of an asteroid the size of Mount Everest, the impact of which resulted in their extinction. Had that asteroid's path through the inner solar system been even slightly different, it would have missed the Earth entirely, as countless asteroids do all the time, and perhaps might have been shunted off to some other nether region of the solar system. But as every assassin knows, it's the hits, not the misses, that change history, and this bull's-eye hit our planet's reset button as only a large asteroid can. In the aftermath, the evolutionary course of life took a dramatic new heading—

away from reptiles and toward mammals; in other words, toward us. If not for that asteroid, we wouldn't be here. But tell that to the dinosaurs.

We are connected to the cosmos in an abundance of ways, connections that may nurture, destroy, or redirect life. This book is about those connections—both known and speculative—how they have touched us in the past and how they may touch us again in the future, assuming we have one that extends for at least several more millennia. Knowing how astronomical influences have shaped our world and enabled the human race to evolve and flourish gives us a unique perspective on the nature and direction of life on Earth and the possibility of life on other planets. They are also sobering reminders of our technological limitations, for until we can harness the energy of the stars, physically relocate ourselves in the Galaxy, or manipulate the space-time continuum, we must humble ourselves on the altar of the universe. Truth be told, there would be no surviving the radiation of a nearby supernova or the impact of a mountain-size asteroid. But knowing that such forces exist—forces that could level 10,000 years' worth of accumulated civilization—ought to humble us into being better stewards of our planet, which, in case nobody noticed, could exist just fine, and would probably be better off, without us.

CHAPTER 1

TILT-A-WHIRL WORLD

It is one thing to say that the ice was there, quite another to say how it got there. If the origin of mountains is sublimely moot, so is the origin of the ice.
—John McPhee, *Annals of the Former World*

The eastern half of Wisconsin is one giant ice carving—a state carved not out of ice but *by* ice. If you hop on a bicycle east of Madison and pedal along the Glacial Drumlin State Trail that begins in the village of Cottage Grove, you'll see what I mean. The trail follows the old Chicago and Northwestern Railroad bed and runs eastward for some 50 miles through the kind of beautiful scenery Wisconsin is famous for: a series of lush, irregularly shaped hills and ridges interspersed with deep hollows, some of which form modest lakes. Most striking are the lines of humpbacked hills called drumlins, from which the trail gets its name. Elsewhere in the state, farmers have plowed concentric contours around the drumlins—making them all the more eye-popping, especially when viewed from the air.

All of these features are the handiwork of glaciers that rasped back and forth across the eastern and northwestern parts of Wisconsin some 10,000 years ago, near the end of the world's last ice age. The irregular hills, called moraines, consist of loose till scraped up by a glacier and dumped along its side, or "snout." The hollows, called kettles, are moraine counterparts. They formed when large, detached blocks of ice melted, depositing sand and gravel in the space formerly occupied by the ice. Drumlins are akin to wave ripples and testify to how geometrical glacial erosion can be. Their oval shapes were sculpted in the direction the glacial ice was moving.

Wisconsin is not the only place where you can see glacial artwork of one form or another. In his book *Annals of the Former World*, John McPhee points out that as the last ice age was winding down, three-fifths of the world's ice—as much as Antarctica holds today—lay across North America.[1] The advancing ice heaped up the headlands of Cape Cod as well as the hills of Michigan, formed waves of drumlins from Alberta to Quebec, carved out the Finger Lakes of New York, and scoured parallel grooves and curvy striations in rock outcroppings in Central Park. If you look across latitudes north of Wisconsin, from Newfoundland westward across Canada and Alaska, you can bet the hills, hollows, and ridges there were sculpted by ice more than 10,000 years ago. And at least another fifth of the world's ice lay across much of Europe and other locales, so similar glacial features can be found in Ireland, Scotland, Poland, France, Austria, Scandinavia, and Russia. In fact, practically anywhere you look in the world's high latitudes, including those in the Southern Hemisphere (in New Zealand and Patagonia, for example), you will find evidence that giant ice sheets once roamed the Earth.

Today, we take for granted that periodic glaciations were responsible for reshaping the Earth's surface, but a little over a century and a half ago many scientists scoffed at such a notion. The existence of glaciers was not in dispute, but the idea that they could reshape the face of the planet was—in part because of the polarized thinking about geology at the time. Scientists generally preferred to believe that the Earth's mountains and valleys had either been reshaped by catastrophic events of biblical proportions, such as Noah's flood (an idea called *catastrophism*), or that such surface features developed in a uniform manner slowly over time, just as a tree changes its shape and size as it grows (*uniformitarianism*).

But beginning in the mid-nineteenth century, a Swiss naturalist named Louis Agassiz gathered enough evidence to show that glaciers had not only transported rocks over great distances but had been more extensive in the past. This new idea was greeted skeptically by many, but being Swiss, and hence well acquainted with the glaciers in the Alps, Agassiz could be viewed as having the advantage.

Ancient glaciers, he wrote in 1840, had "enfolded entire mountain chains and descended into the plain by great valleys."[2] As proof, he pointed to the "erratic blocks" he found resting on beds of rounded cobbles in the Jura Mountains on the Swiss-French border. The largest cobbles, he noted, were on top; smaller pebbles were in the middle; and the smallest, down to sand, were on the bottom. This arrangement, he said, ruled out the notion that they had been carried there by "rafts" of ice on currents of water, as some noted geologists at the time—Sir Charles Lyell, for one—argued. If the cobbles had been so conveyed, the order of their superposition would have been reversed. The conclusion, then, was irrefutable: the blocks and pebbles could only have been transported there by a massive glacier.

In 1839, Agassiz traveled to the Alps and trekked up the glaciers on the apron of the Matterhorn, on the border between

Switzerland and Italy. From there he hiked up the Aar glacier to the base of the Finsteraarhorn, the highest mountain in the Bernese Oberland. There he learned that a cabin built by a monk in 1827 was now located more than 600 meters (2,000 feet) down the mountain.[3] Agassiz built his own shelter on the Aar glacier and, with the help of some students, pounded a row of stakes in a straight line across the ice. By 1841, he found that the most rapid movement was in the center of the glacier, owing to the friction with the mountain walls, so that after a few years the stakes were arranged in a U shape. Glaciers, he concluded, flowed—not as quickly as rivers, but not unlike them either.

Given findings such as these, objections to glacial theory soon melted away. Agassiz had synthesized a convincing picture of how glaciers had shaped the Earth. Glaciers not only moved, but in the geologic past they had advanced and spread over areas that today were free of glaciers, save for their telltale erosive traces. Agassiz found evidence of glaciation throughout Europe, the British Isles, and, eventually, North America. Today, he is considered the father of glaciology.

Agassiz gave the world the ice age concept, but during the last part of the nineteenth century it dawned on scientists that there wasn't just one ice age in the Earth's history, but many, with warm periods in between. In the wake of these revelations, there followed a laborious quest on two fronts: to determine where and how often glaciation occurred, and to search for the cause of the Earth's apparently cyclical ice ages. The first would impel scientists to dig into the suggestive scars left by the abrasive passage of prehistoric glaciers, while the second would lead a handful of theorists to look up at the sky and fault the stars—more specifically, the various complex geometries involved in our planet's orbit around the Sun.

The first quest naturally drew sober, empirically minded geologists and meteorologists, but the latter quest was not without its champions. Two early visionaries were French mathematician Joseph Adhemar and Scottish scientist James Croll. In his 1842 book, *Revolutions of the Sea*, Adhemar was the first to suggest that ice ages could be tied to changes in the Earth's orbit, specifically its eccentricity. Nothing wrong with that, of course, but the rest of his theory was just too elaborate to believe. Adhemar assumed that when the Earth was in the grip of an ice age, the polar ice cap most affected by the deep freeze would extend down to the sea floor, whereupon the ice cap would grow like Pinocchio's nose until it reached a spectacular altitude of 100 kilometers (62 miles). Over the following 10,000 years, as the wobble of Earth's axis (called precession) slowly brought the frozen pole back into the more direct rays of the Sun, the warming ocean currents would eat away at the base of the polar ice cap, eventually causing it to topple catastrophically into the ocean. This, in turn, would create a massive and destructive tidal wave, a real end-of-the-world event. Obviously, Adhemar was a catastrophist.[4]

Not surprising, his theory was almost immediately quashed, but it nevertheless inspired Croll to come up with his own, less hypothetical, variation. Croll, a self-taught scientist of no-nonsense Scottish descent, summed up his idea in a number of books, one of which he titled *Climate and Cosmology* (1885). Using orbital calculations devised by French astronomer Urbain Jean Joseph Le Verrier and Italian-French astronomer and mathematician Joseph Louis Lagrange, Croll extended the calculations for the maximum eccentricity of the Earth's orbit over the course of four million years. Croll proposed that changes in eccentricity—how much the orbit could get out-of-round over time—were significant enough to account for all the extremes in climate. His theory, which also factored in the axial wobble of precession, predicted that when Earth's orbit became more elliptically shaped than circular, as it was known to do, winters were coldest in the hemisphere that was inclined away from the Sun at its most distant point, called aphelion.[5] Additionally, the increased snowfall during these periods would increase the planet's reflectivity, or albedo, making the planet colder still. Thus, ice ages would occur in cycles of 26,000 years (the precessional cycle), alternating between the Northern and Southern hemispheres, and last for approximately 10,000 years. Like Adhemar, Croll's ideas were not well received by many scientists at the time, though Lyell (a dyed-in-the-wool uniformitarianist) and Charles Darwin found favor with them.

The question of what caused ice ages raised its snappish head again in September 1936 during the third meeting of the International Quaternary Association in Vienna. INQUA was founded in 1928 by a group of scientists seeking to improve understanding of environmental change during the glacial ages.* Most of the scientists attending the convention were geologists and geographers, with the exception of a soft-spoken Serbian mathematician named Milutin Milankovitch (the Serbian spelling of his name is Milankovic´). Like Adhemar and Croll, he, too, had developed a theory that attributed the ice ages to astronomical forces, only his theory was far more extensive and backed by mathematical proofs.

Nevertheless, his ideas had been ignored or overlooked by most of his fellow scientists. Geologists and climatologists argued that astronomical forces were too negligible to have any effect on the climate. Besides, the ice ages appeared to be contemporaneous in both the Northern and Southern hemispheres—not alternating

---

*During that same meeting, the association formally changed its name to *Internationale Quartärvereinigung*, or the International Association for Quaternary Research. The acronym INQUA, proposed by one of the members, has been retained ever since.

Figure 1. Portrait of Milutin Milankovitch by Paja Jovanovic, 1943. Courtesy of John Imbrie, Brown University.

between one hemisphere and then the other. As such, the cause of the ice age could not be ascribed solely to changes in the tilt of Earth's axis or the shape of its orbit.

But Milankovitch stood by his theory, including at the Vienna meeting. Not being a geologist or a geographer, he normally would not have attended such a gathering, but because of his long labor on the ice age problem, he had been invited. Besides, he wanted to see his "beloved Vienna" again, where he had been educated as a young man. And he was very interested in hearing the association's opening scientific paper, given by none other than Albrecht Penck, Germany's greatest geographer, not to mention the INQUA's founder and honorary president.

Penck was a geographic legend in his own time. He had held professorships at the Universities of Vienna and Berlin and for sixteen years had been the director of the oceanography and geography institutes in Berlin. The INQUA established a career award in his name, and an oceanographic research ship named for him still operates out of the Baltic Sea Research Institute in Germany. But Penck made science history in 1909 when he and an Austrian meteorologist and geographer named Eduard Brückner produced the first definitive proof of repeated growths of Alpine glaciers, thus confirming Agassiz's original findings. The men studied the kettles, moraines, river gravels, and terraces in the northern foothill valleys of the Alps and identified four periods of glaciations for the region: Günz (the oldest), Mindel, Riss, and Würm (the most recent), so named for the Bavarian streams in which the moraines and gravels of particular episodes were well represented. Penck and Brückner's massive three-volume work, *Die Alpen im Eiszeitalter* (The Alps in the Ice Age), served as the Bible of glacial geology. The discovery also transformed geology and climatology, opening new fields of research and providing scientists with a more dynamic view of how the Earth's surface changed over tens of millions of years.

Twenty-seven years downstream from his and Brückner's landmark discovery, Penck's giant reputation was only enhanced by his resonant voice and eloquence. He was at the time in his life when younger scientists sought to curry favor with the great man and to hear him relive past discoveries. Penck was only too happy to oblige. His presentation to the Quaternary Association, titled "The Climate of the Ice Ages," was basically a recounting of his work with Brückner and their discovery of the four periods of glaciation. Penck reminded the audience that each of the four ice ages not only had had cool summer years but low average yearly temperatures as well. Nevertheless, he flatly rejected the idea that cooler summers contributed to increases in the Earth's surface coverage of snow and ice.

Sitting in the audience, Milankovitch knew that *his* work demonstrated otherwise. In 1933, he developed calculations that linked the size and altitude of snow caps to changes in hemispheric temperature. Essentially, the calculations showed that if a snowline descends to a lower level, the snowcap, in turn, enlarges. Like a giant mirror, snow reflects solar radiation back into space, cooling the hemisphere and causing the snowline to drop even lower. These fluctuations of the snow caps could be tied into changes in the inclination of Earth's axis, which changed the amount of solar radiation reaching a given area of the surface (a factor called *insolation*). That same year, Milankovitch presented his findings, "New Results of the Astronomical Theory of Climatic Changes," to the Royal Serbian Academy of Sciences.

Three years later in Vienna, by Milankovitch's own account, Penck's speech was a great success. After the applause subsided, the presiding member of the session asked if the audience had any questions or comments. Milankovitch abruptly decided to go back on his resolution of being only an observer. A complete opposite in personality to the forceful Penck, he hesitantly rose and in a soft, halting voice that only a few sitting nearby could hear, stated that his calculations, too, showed that during the main phases of the ice ages not only the summer but the average yearly temperatures were well below those of the present. But, Milankovitch emphasized, it was these conditions that contributed to the advancing ice line. The primary cause of those lower temperatures, he went on to say, was a phase in which the Earth's axis was more straight up and down—in other words, more perpendicular with respect to the plane of the solar system. "With the straightening of the axis," he continued, "the yearly radiation of polar and adjacent regions is reduced—these being where the ice ages took place."

In the notes from his autobiography, assembled after his death by his son Vasko, Milankovitch recalls what happened next: "When Penck returned to the rostrum it was obvious that he was annoyed: 'Astronomical theories do not give us the answer to the Ice Ages!...I studied Milankovitch's theory and have rejected it. ...The ice ages are not a product of changes of astronomical elements but of periodic changes in the thermal power of the Sun.'"[6]

Penck turned from the rostrum and, to Milankovitch's eye, was "visibly upset and exhausted."

Aside from Penck's tirade, there is a curious aspect to this exchange. There is little evidence, other than Milankovitch's own account, to suggest that Penck openly stated, or believed, that ice ages were due to changes in the "thermal power of the Sun." One of the more favored theories at the time was that tectonic uplift of mountain ranges or other surface features would alter the circulation of ocean currents and wind patterns, thus giving rise to temperature swings such as ice ages. Moreover, to state emphatically that the ice ages are not a product of astronomical forces and then, in the same breath, to cite one that is, seems contradictory. In any case, it suffices to say that Milankovitch's theory was not one of which Penck openly approved.

Milankovitch thought that he had made an "unforgivable guff" and immediately withdrew from the meeting. It was, he also found out, Penck's seventy-eighth birthday, which only increased his feeling that he had committed a terrible faux pas. Then, two years later, in a complete and surprising reversal, Penck wrote a scientific paper in which he *praised* Milankovitch's theory as the "real canon of secular insolation" of the Earth during the past 600,000 years.[7] As gratifying as such an endorsement must have been, it was not enough to help Milankovitch avoid the slings and arrows of outrageous criticism. His astronomical theory of climate change would not find favor for another thirty years.

Little in Milankovitch's background prepared him for the world of academic disputes. Born in 1879 in the rural village of Dalj, Serbia, he was the eldest of four sons. In his autobiography, Milankovitch describes himself as the weakest of his siblings who cared not for sports or physical exertion. "While my brothers played in the garden, skated on the ice, exercised on gymnastic bars, and generally developed their bodies, I sat in my little room, building little houses, reading books, and solving puzzles," he wrote.[8] His exertions were clearly in other directions. During his early school years, he marshaled his analytic skills by excelling in mathematics. Though his father wanted him to study agricultural science and eventually take over the family estate, Milankovitch had no desire to be a farmer. In 1896, at age seventeen, he attended the Technical High School of Vienna, where he pursued a five-year course in civil engineering. After obtaining his doctorate, he briefly served as chief engineer for a construction company. His stint as engineer did not last long, however. In 1909, he was offered, and accepted, a faculty position in applied mathematics at the University of Belgrade. Perhaps it is a testament to his personal inertia that he held this position for the rest of his life. All he wanted, he said, was to be able to live and work in peace "without having to rush."

Milankovitch was interested in solving puzzles that had yet to be solved, but not just any puzzles. He saw little point in working on minor problems in already cultivated fields. What he really wanted to do was solve a major scientific problem that would not only break new ground but, frankly, write his name into the science history books. He found such a challenge in the riddle of what caused the ice ages.

One August evening in 1913, as he was admiring the Moon over the Danube, Milankovitch mused that though scientists had mapped and measured its visible features, they still did not know its surface temperature or what changes it experienced. This led him to consider how long-term changes in the orbit of a planet, particularly the Earth, could have climatological consequences. Milankovitch realized that if he could find a relationship between the Sun and surface temperature, he could "numerically trace the effects of the Sun's rays on the atmosphere and on the Earth's surface." By 1914, he had published six papers on what he called his "mathematical theory of climate."

Over the following years, Milankovitch built on the work of Croll, charting the onset and retreat of the ice ages during the Pleistocene epoch, which began 1.8 million years ago and ended when the last ice age died out some 10,000 years ago. Milankovitch formulated a model that calculated latitudinal differences in solar radiation for a 600,000-year period prior to 1800 CE. When he factored in celestial mechanics, he realized that the differences could be explained by a combination of three primary astronomical cycles: variations in our planet's inclined angle to the Sun; a wobbling motion inherent in the Earth's axis; and a gradual change in the shape of the Earth's orbit around the Sun. Over tens of thousands of years these factors worked together, and sometimes against each other, to force-change the Earth's climate.

But as Milankovitch saw in Vienna, his efforts were virtually ignored. Geologists didn't trust astronomical-based theories. And astronomers who specialized in celestial mechanics were more interested in applying these disciplines to studying the motions of other planets, not climate on the Earth. Thus, Milankovitch's ideas were either avoided or simply went unnoticed. He was not unfamiliar with academic shunning; he was, after all, a contemporary and friend[9] of Alfred Wegener, the German geologist who, in 1912, proposed the theory of continental drift, a theory that had been deemed "utter, damned rot!" by no less than the president of the American Philosophical Society.[10] Wegener died during a 1930 expedition to Greenland, twenty-five years before continental drift would become a respected theory and then, by the 1960s, confirmed. Milankovitch, too, would live out his life with the reputation of being the ward of an orphan theory and then die in 1958, some fifteen years before the theory gained acceptance.

Looking back, it's easy to see why Milankovitch had trouble convincing others. On a day-to-day basis, the weather at any location on Earth is a sum result of wind, humidity, temperature, air masses, local topography, and so on. But climate is a much more complicated beast than weather, involving long-term trends and changes in factors such as ocean and wind currents. Milankovitch, however, believed that although there may be factors that govern temperature in the short term, over periods greater than 10,000 years his mechanism—the Earth's position with respect to the Sun—was the only one that mattered. Heat came from the Sun in a steady stream, Milankovitch reasoned (although this is not entirely the case, as I show in the next chapter). If *anything* varied, it was the Earth's orientation to the Sun, and there was physical and mathematical evidence to prove it.

Milankovitch did not merely see the Earth and its sediments; he saw the Earth in space and in motion around the Sun over the course of millions of years. It took uncanny vision to step off the Earth and look back from a distance of 100 million miles and watch the cogs turn, then forge a mathematical connection between those motions and the boulders and cobblestones on the Jura Mountains. It was the same kind of vision possessed by people like Agassiz, Adhemar, Croll, and Wegener, some of whom paid a high price to see worlds, possibilities, and connections that others could, or would, not.

To grasp what Milankovitch saw, we need to do the same thing: imagine the Earth rotating about its axis and revolving around the Sun over geologic periods of time. It might help to envision a tilt-a-whirl carnival ride, a spinning top, or a toy gyroscope as analogs. The point to bear in mind is that all aspects of Earth's orbit and axial orientation are constantly changing—though in our brief life spans those changes go unnoticed.

In Milankovitch's scheme, one of the most dominant climate control mechanisms is Earth's obliquity, or the angle between the axis around which it rotates and the plane of the solar system. As we all learned (or should have learned) in geography class, our trusty north-south axis is tilted 23.4° away from being exactly perpendicular to the plane of the solar system. Since the axis essentially points toward the same direction in space as Earth revolves around the Sun, each pole is angled toward the Sun for part of the

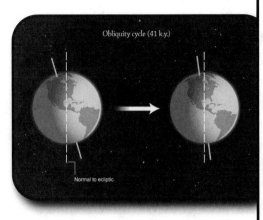

Figure 2. Over a cycle of about 41,000 years, the Earth's axis oscillates between a minimum of 22.1° (right) and a maximum of 24.5° (left). The Earth's axis is currently near the middle of this swing, at an angle of 23.4°. In another 10,000 years, the inclination will once again reach its minimum point. Illustration by Nik Spencer.

year and angled away from it for the other part. It is this obliquity that causes the seasons.

But what's not often mentioned in beginning geography class is that Earth's axis has not always been, and will not always remain, tilted by 23.4°. Over a cycle of some 41,000 years, its inclination jockeys between a minimum of 22.1° and a maximum of 24.5°.* It's a small amount, but it makes a big difference in insolation, the intensity of incoming solar radiation measured at a planet's surface or at the top of its atmosphere (if it has one). A bigger inclination results in more severe seasons at the middle and upper latitudes for both hemispheres. Winters would be colder, and summers hotter. A smaller inclination, on the other hand, translates into less severe seasons at these latitudes—milder winters and cooler summers. As Milankovitch argued, in general a high-tilt phase tends to melt ice sheets that may have formed

---

*We are fortunate, indeed, that the difference amounts to only 2.4°. Had the primordial Earth not acquired a satellite with the mass of the Moon, the obliquity of the axis would vary chaotically, ranging from 50° to 85°. This would have drastically affected Earth's climate, not to mention the evolution of life.

during a low-tilt phase, bringing an end to encroaching glaciers at subpolar latitudes. The effect is seen in the chronology of the last ice age, which peaked some 18,000 years ago, after the inclination had increased from its 22.1° minimum to 23.4°. Over the next 10,000 years, as the axis reached its maximum of 24.5°, the ice began to rapidly retreat. Since then, the inclination has waned until today we find ourselves once again at 23.4°. In another 10,000 years, the inclination will once again reach its minimum point. Does that spell another ice age? Read on.

The second orbital climate force is the shape of Earth's orbit. Like the other planets in the solar system, Earth's orbit is slightly oval shaped, or elliptical. As every geometry maven knows, an ellipse has two focal points located along the major axis, or the long part of the oval. A measure of an ellipse's departure from circularity—as the brain-endowed scarecrow from *The Wizard of Oz* might recite—"is given by the ratio of the distance between the two focal points to the length of the major axis." This measure is called *eccentricity*. Simply put, the more elliptical an orbit is, the greater its eccentricity. A computer-generated circle would have an eccentricity of 0, whereas the best we humans could muster without a compass might have an eccentricity of about 0.05, which would still be pretty darn good. More stretched-out circles can have eccentricities up to, but not quite, 1. For an extreme example, Halley's comet has an eccentricity of 0.9673, which resembles something like an elongated rubber band.

Earth's present eccentricity is a modest 0.017, which is very close to a computer-generated circle, but not precisely. The slight departure from perfection translates into a difference of just 5 million kilometers (3 million miles) between the point when Earth is closest to the Sun during the year—*perihelion*—and when it is at its greatest distance—*aphelion*. The difference amounts to

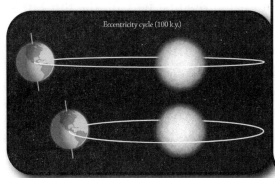

Figure 3. The gravitational force exerted by Jupiter and the other planets causes the Earth's eccentricity to vary every 100,000 years, in which the orbit changes from nearly circular (below) to slightly oval (above). Illustration by Nik Spencer.

only about a 6 or 7 percent increase in insolation between aphelion and perihelion.

But the Earth's orbit wasn't always so circular, nor will it remain so. The gravitational force exerted by Jupiter and the other planets causes Earth's eccentricity to vary roughly every 100,000 years, in which the orbit changes from nearly circular (an eccentricity of 0.005) to slightly oval (0.028). There is also a larger, more pronounced cycle every 413,000 years, during which the Earth's orbit becomes more significantly out-of-round, reaching an eccentricity close to 0.06. This means the amount of insolation received at perihelion can be 20 to 30 percent greater than it is today.

Finally, the cycle of *precession* adds another level of complexity to this already baroque armature. Over eons, the gravitational tug on the Earth's equatorial bulge by the Moon and, to a lesser extent, the Sun, have superimposed a toplike gyration on its axis. As a consequence, the vernal equinox, the point in the sky where the plane of the ecliptic and the celestial equator cross, regresses about 50 arcseconds per year.[11] One complete gyration takes 25,765 years, during which the axis circumscribes a great circle among the stars. Currently, the northern axis points toward Polaris, the North Star, but during the first half of the third millennium BCE, around the time the pyramids of Egypt were being built, the axis pointed toward another star, Thuban, in the constellation Draco. Thuban, then, served as the Egyptians' unfailing "North Star." Some 12,000 years from now, the axis will point near the bright star Vega in the constellation Lyra.

In addition to the precession of the Earth's axis, its orbit, too,

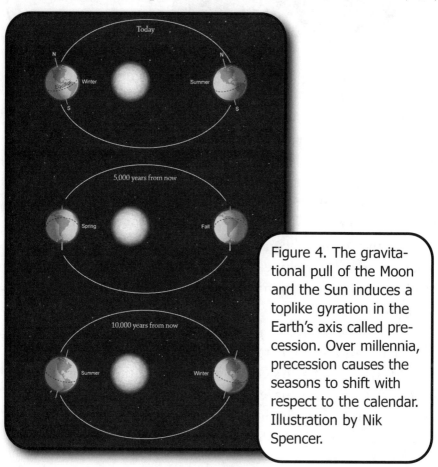

Figure 4. The gravitational pull of the Moon and the Sun induces a toplike gyration in the Earth's axis called precession. Over millennia, precession causes the seasons to shift with respect to the calendar. Illustration by Nik Spencer.

precesses in space due to gravitational perturbations exerted by the other planets, primarily Jupiter and Saturn. In this case, the thing doing the precessing is the line drawn through the Earth's orbit from the point of perihelion to the point of aphelion, called *the line of apsides*. (Essentially, the line of apsides is the major axis of an elliptical orbit.) The speed of this rotation amounts to 47 arcseconds per century, where one arcsecond is one-3,600th of a degree on the sky, or about the size a quarter would appear seen at a distance of 5 kilometers (3 miles).[12] In this cycle, the Earth's perihelion point completes one revolution with respect to the vernal equinox every 21,000 years.

Over time, on the order of thousands of years, these precessional motions cause the seasons to shift through the calendar. More important to Milankovitch's theory, they change the orientation of the Earth's axis with respect to the moments of perihelion and aphelion, with climatic consequences. At our present time in history, it is no coincidence that the winter and summer solstices occur near the dates of aphelion and perihelion. That is simply a legacy handed down by our sky-watching ancestors who noticed when the Sun had reached its northernmost and southernmost extremes in the sky for the year. But notice that the dates for the onset of winter and summer do not *exactly* correspond to the moments of perihelion and aphelion. The winter solstice generally occurs on or about December 21, while the summer solstice occurs on or about June 21. Perihelion, however, occurs on or about January 3, while aphelion occurs on or about July 3. On average, there are 13 to 15 days' difference between the days of the winter and summer solstices and the occurrence of perihelion and aphelion.

Hence, although the dates of the winter and summer solstices remain pretty much the same every year, the moment of perihelion and aphelion, to which the solstices are related, creep forward a little and the calendar lags behind. Every sixty years, the date of perihelion arrives progressively later by about one full day. In the 1860s, for example, the Sun reached perihelion around the first of January. By the year 2078, perihelion will occur on or around January 4. Looking ahead 5,000 years, perihelion will occur around the spring equinox and aphelion during the autumnal equinox. In 10,000 years, perihelion will occur at the summer solstice, when the Northern Hemisphere will be more inclined toward the Sun and the Southern Hemisphere away from it. These circumstances are the reverse of what we have today.*

Does that mean warmer summers for northerners in 10,000 years? Not necessarily. Insolation may be greater, but it must be remembered that there isn't a strict one-to-one relationship between radiation values and temperature. The difference between the closest and farthest points from the Sun amounts to only 3 percent; hence other factors like cloud and snow cover and ocean currents can have a greater effect on surface temperature. Further, the mechanical aspects of an elliptical orbit must be kept in mind. Kepler's second law of planetary motion dictates that a line connecting any orbiting planet to the Sun must sweep out equal areas in equal times. At aphelion, when it is farthest from the Sun (and the solar system's center of gravity), a planet orbits more slowly. But because of the greater Sun-planet distance, a lot of area is nonetheless swept out between any two points. To sweep out an equivalent area at perihelion (when it is nearest to the solar system's center of gravity), a planet must move more rapidly in its orbit. So what this means is that in 10,000 years' time, the Northern Hemisphere's summer season will be shorter and its winter season longer. The Southern Hemisphere, of course, will

---

*The actual dates of those future perihelions and aphelions are not important, since our Gregorian calendar will be obsolete in 10,000 years.

experience the reverse: longer summers and shorter winters. Moreover, the Earth's axis will, by then, be at its minimum inclination, and the moderation of insolation over time could spell milder winters and cooler summers.

To be sure, it's a complicated business and difficult to describe in a few sentences. But looking simply at the orbital factors that can influence climate trends, we can state that, in general, latitudinal insolation for Hemisphere X is least when obliquity is at its most extreme tilt and precession is such that perihelion occurs in that hemisphere's winter. Conversely, insolation is greater for Hemisphere Y when obliquity is at maximum and perihelion occurs during that hemisphere's summer. Such were the conditions 10,000 years ago in the Northern Hemisphere, when the Earth was emerging from its last ice age. And this is how Milankovitch was able to explain these planetary-scale changes in climate: by looking at our place in the solar system.

By itself, eccentricity has almost no effect on climate, but, as inferred by Milankovitch, when coupled with extremes in obliquity and the right moment of precession, you might get something much more noticeable. It looks as if there has to be a special combination of cycle moments—a congruence, for example, of minimum and maximum events—to yield a big change in climate. Such was the case 23 million years ago, when minimal obliquity coincided with low eccentricity, resulting in a period of reduced seasonal extremes. This, in turn, led to the buildup of vast ice sheets across Antarctica, reversing what had been a long stretch of natural global warming. And working with tiny fossils in deep-sea sediments, researchers have unearthed detailed records of past temperature changes over the past 14 million years, correlating to changes in—you guessed it—obliquity, precession, and eccentricity.

The Earth is not alone in having its climate subject to cycles of orbital forcing. Milankovitch's theory of thermal and radiation effects can also be applied to other planets, particularly Mars. In his work *Théorie mathématique des phénomènes thermiques produits par la radiation solaire*, published in 1920, Milankovitch concluded that since the amount of solar radiation reaching the Earth is 2.3 times greater than that reaching Mars, the Red Planet could probably not sustain life.[13] On Mars, the tilt of the rotational axis can get much higher than that on the Earth, leading to extreme warming of the Martian polar caps. Water ice frozen at the polar caps then evaporates directly into vapor, which migrates to the cooler lower latitudes and freezes again as ice-rich deposits on the surface. Orbiting spacecraft, such as the European Space Agency's Mars Express and NASA's Mars Global Surveyor, have gathered a multitude of evidence supporting the case for surface and subsurface ice—even glaciers—on Mars.[14] Mission cameras have imaged tonguelike features, debris aprons, and "accumulation areas" that have been interpreted as the remnants of glaciers now mantled in dust. Elsewhere on the planet, geologists have identified what they think are glacial moraines and lines of debris in valleys caused by flowing ice. Since ice phases directly to vapor on Mars, the glacial features and debris lines remain intact, forming sort of a geologic afterimage on the planet's surface.

Despite the success of Milankovitch's theory in many arenas, it is not without flaws—some of which are fairly serious. One of these is known as the "100,000-year problem." Because changes in eccentricity during the 100,000-year cycle, when the orbit is nearly circular, have a much smaller impact on climate than obliquity and precession, it should produce the weakest climate signal. However, observations show just the opposite. During the last one million years, the dominant climate signal *is* the 100,000-

year cycle. Conversely, the 400,000-year eccentricity variations, when Earth's orbit is most out-of-round, should produce a strong climate signal, and yet it has so far gone undetected in the paleo-climate records. Some scientists assert that there may be other climate "feedbacks" that enhance the 100,000-year cycle, such as atmospheric carbon dioxide or ice volume or even interplanetary dust. Still others claim that the climate record may not be long enough to establish a statistically significant relationship between climate and eccentricity. It's even possible that unknown chaotic influences may play a role in the Earth's climate, not unlike the nonlinear forces that shape and drive the turbulent eddies of a stream. All of this, of course, remains to be seen.

The relationship between the waxing and waning of the great continental ice sheets and changes in the Earth's orbit has been a long-standing problem in Earth science. That ice and climate are connected to resulting variations in local sunshine is beyond doubt, but how prominent a role do Milankovitch cycles play in the bigger picture of ice ages? Toward that end, one atmospheric scientist has gone so far as to publish a paper titled "In Defense of Milankovitch." Gerard Roe, of the University of Washington in Seattle, says the point he wants to make is how changes in local sunshine, due to orbital factors, could affect the waxing and waning of the planet's great continental ice sheets. That there is some connection is beyond doubt. The problem, Roe says, is to sort out all the physics. In his paper, which appeared in *Geophysical Research Letters* in 2006, Roe notes that changes in insolation are linked to the rate of change in the global volume of ice.[15] That rate of change, he says, is more important than focusing on the absolute global volume of ice. "If you turn the summer sunshine up locally, the first thing that happens is that ice melts," he says. "So there is a much more direct physical connection between summer sunshine and ice *melting* than there is between summer sunshine and ice *volume.*" But the general question of how orbital variations affect the climate of different regions and paleoclimate proxies, says Roe, should be separated from the specific question of the cause of changes in the extent of ice sheets that are orbitally induced.

It is true, says Roe, that to some degree everything influences everything else, but the dominant influences remain elusive. "Every aspect of Earth's climate is affected by orbital variations," he says. "The Sun is obviously the ultimate energy source for the great heat-engine that drives the circulations of the atmosphere and ocean fluid envelopes. We know, for example, that major components of the climate system like the Asian monsoon, El Niño, and the over-turning of the ocean circulation are all impacted. Teasing out the links remains one of the great challenges in climate science."[16]

Like many misunderstood scientific visionaries, Milankovitch was certainly on to something when practically everyone else thought he was not. While there are still other unresolved issues with Milankovitch's ideas, the theory as a whole has largely turned out to be a success. Paleoclimatologists are now starting to find more favor with the notion of orbital climate forcing, including how it influences tropical climates and in the effects of trace gases, such as carbon dioxide and methane, on ice volume. Fortunately, orbital influences on climate take many thousands of years before they are felt, plenty of time for intelligent future inhabitants to prepare for any Milankovitch-style climate changes to come. However, there is yet another astronomical climate force that needs to be considered, one that could have more immediate effects: the Sun itself.

CHAPTER 2

AN IMPERFECT SUN

As people and as scientists, we have always wanted the Sun to be better than other stars and better than it really is.
—Jack Eddy, "The Case of the Missing Sunspots,"
Scientific American, May 1977

The Chamonix Valley in the French Alps is a verdant flat-bottomed ravine some 24 kilometers (15 miles) long but roughly a kilometer (half a mile) wide at its greatest breadth. Nestled as it is between the soaring flanks of lofty mountains—dominated to the south by the summits of Mont Blanc and the Aiguille des Grands Charmoz—it is an idyllic mélange of what is best about pastoral, alpine countryside. The valley's namesake is the town of Chamonix, which lies along the cold, gushing, glacial meltwaters of the Arve River. Everywhere you look from this vantage point you see lofty, snow-covered mountains, expanses of fir and larch forests, and, in spring and summer, sprawling meadows blanketed with a tapestry of colorful wildflowers of every variety. Southwest of Chamonix the eye is irresistibly drawn to a conspicuous glacier called Les Bossons draped across the northwestern flank of Mont Blanc, its snout bristling with jagged ice pinnacles called seracs. In the slanting late-afternoon light, the ice blushes to a hue similar to that of rose quartz. Take away the swarming tourists, screechy motorbikes, and throbbing techno bars, and Chamonix would be a true Shangri-la.

But an even more spectacular sight is to be found up the north-eastern flank of Mont Blanc, from a village called Montenvers. To get there, one must either hike (about 2.5 hours one way) or take a 20-minute rack-rail train ride. Either way, the trip is worth it. Montenvers overlooks one of the most spectacular glaciers in the Alps: the Mer de Glace, or "sea of ice." From space, the 12-kilometer-long glacier looks like a ribbed serpent. The "ribs" are Forbes bands, transverse alternating bands of light and dark ice that are the result of different dirt and snow deposition in summer and winter. They bend with the ice flow of the Mer de Glace and as such are, in a sense, time stamps of the glacier's movement.

Above Montenvers, the glacier stretches nearly into the sky. The upper regions, which curve left then right up the glacial valley before vanishing behind jagged peaks, are brilliant white, while directly below, closer to the terminus, the ice is mantled in a chalky, rocky soil. If you look carefully, you can see fissures here and there in the crust that vanish into a blush of deep blue ice.

The view is gripping, but it is the glacier itself that beguiles. It is not unusual, I'm told, for those who see it for the first time to feel a tingle of unease and for the hairs to rise on the backs of their necks. And when someone, a guide perhaps, remarks that it was in this locale that author Mary Shelley depicted the disturbing confrontation between Victor Frankenstein and his monster, you realize you are regarding something that prods at the primal brain.

Figure 5. The glacier Mer de Glace as seen in July 2005 from Montenvers, France. Courtesy of Alexandra Witze.

For those agile enough, a rough path leads down to the Mer de Glace itself (or you can take the cable car). You can even go inside the glacier via an artificially hollowed ice grotto. It's cold and dark within, and disoriented tourists bump around as if in a dream. There are ice furniture and animal carvings, and even a gift stand where (for a small fee) you can have your photo taken next to a huge Saint Bernard with the traditional keg hanging from its neck. Everywhere is the sound of dripping water, and you can, in places, let it drip-drop upon your tongue. The water has no taste to speak of. The only word that comes to mind is *ancient*.

The ice along the grotto's entrance is a spectacular spectral blue, with bits of soil and gravel suspended clearly within. This material is likely the detritus of mountains that were rasped by the glacier millions of years ago, settling through the glacier like pearls through molasses, and finally revealed in the wall of the ice grotto. It's a reminder that, though you can't see it, the glacier is continuously moving, dragged inexorably by gravity toward the Chamonix Valley at a rate of about one centimeter (less than half an inch) per hour. That's pretty rapid as glaciers go; the movement may be discerned from atop the glacier when snow and soil on the surface suddenly topple together and roll downgrade on their own, as if nudged by an unseen foot.

Still, the glacier is not likely to reach the valley anytime soon, if ever. Global warming is seeing to that. Like most glaciers in the world, the Mer de Glace is retreating. Perhaps that is a good thing, for to look at it now—as broad, deep, and majestic as it is—it is difficult to believe that it once made life a misery for folk in the Chamonix Valley.

Today, the Mer de Glace is not visible from the town of Chamonix. But centuries ago, it would have riveted your attention. For anyone who saw it for the first time, the Mer de Glace was said to be "terrible and frightening to look on."[1] The year was 1644, the month of May, a time when the summer Sun should have been warming the countryside, but here it always seemed like winter, particularly with the glacier looming and advancing, at times, "by over a musket shot every day." Its thick marbled tongue slumped 925 meters (3,000 feet) from an extended cleft at the northeastern base of Mont Blanc and lolled onto the floor of the valley, dwarfing houses, livestock, and people. Behind it, the serrated spires of Les Drus jutted into the mist like fangs. For unsuspecting visitors, the "great and horrible" glacier stirred a nameless and primordial fear.

Seen with a more aesthetic eye, the glacier—also called the Glacier des Bois for the abandoned hamlet that lies nearest to it—looked like an enormous glaucous cascade whose turbulent white-water waves had been abruptly frozen in place by some uncanny power. For those who lived in its shadow, however, it was neither fixed nor frozen. It was alive and quite possibly possessed. You could actually see it edging closer, relentlessly plowing over trees and fertile meadows. At night, you could hear it: the steady, coarse sibilance of stone being ceaselessly ground beneath the ice like the devil's own mortar and pestle. This was often punctuated by thunderous booms that echoed throughout the valley whenever a house-size serac suddenly calved from the glacier and crashed to the rocks below. For more than sixty years, the Mer de Glace had menaced the inhabitants in the Chamonix and adjacent Arve valleys. Many Chamoniards had lived out their lives in its shadow, never knowing a time when it wasn't hanging over them like an icy sword of Damocles.

Its advances of late, however, were more pronounced and menacing. The large crosses driven into the ground at the glacier's snout, as a way of divinely stopping its advance, never stood upright for long. Everyone in Chamonix knew what fates the Mont Blanc glaciers had handed the hamlets of Le Bois, Le Châtelard, Bonanay, and especially Les Rousier. All had stood for centuries, and now, in a matter of a few years, they were either buried under the ice or abandoned. Dozens of houses had been engulfed, hundreds of acres lost, and people swept away in sudden floods or buried by avalanches. Good arable land had been covered over by ice and glacial meltwater. But Les Rousier! On June 22, 1610, the leading edge of the adjacent Argentière glacier fell upon Les Rousier, sweeping away eight houses and barns and burying nearly 100 acres *in a single day*.

Between 1628 and 1640, the embattled valley saw at least a third of the cultivatable land lost through avalanches, snowfall, meltwater, and the continued advance of the glaciers. No one dared sow in the autumn because by winter the fields would be covered with snow. An arbitration report made on May 28, 1642, stated that many inhabitants were suffering from malnutrition and ill health.[2] The people living there, the report declared, "are dark and wretched and seem only half alive." It was said by the locals that the land around was bewitched and that "evil spells" were at work among the glaciers. By 1643, Jean Duffong, the administrator of the priory church of Chamonix, reported that the glaciers were continuing to press into the valley, distressing both refugees and their neighbors, all of whom had to deal with diminished crop yields and frequent and sudden flooding.

Though no one realized it at the time, by 1644 the Mer de Glace had reached its greatest length.[3] The glacier was so swollen that it threatened to dam up the Arve. If that happened, water would amass into a great lake behind the Mer de Glace, and, as so often was the case, if the glacial dam suddenly gave way, it could completely flood the Chamonix Valley. In near panic, the syndics of Chamonix appealed to Charles August de Sales, the bishop of Geneva, asking him if the hated glacier was God's punishment for their sins. They pleaded with the bishop for divine intervention. De Sales promised to help, and that June he led a procession of three hundred people up to the snout of the Mer de Glace, as well as those of the glaciers Argentière, Le Tour, and Les Bossons, sprinkled each with holy water, and ritually blessed them.

The bishop's blessing apparently worked, because over the next nineteen years the glaciers gradually retreated. Unfortunately, the providential solution was a brief one because by 1663 the glaciers were threatening again. This time a series of exorcisms were performed. Jean d'Arenthon, de Sales's successor, was summoned to Chamonix at least twice, in October 1664 and again in August 1690. After d'Arenthon's first exorcism, the Chamoniards reported that the glaciers had retreated "more than eighty paces"; after his second, they had withdrawn an eighth of a league, or about 500 meters (1,640 feet), from where they had previously wreaked havoc.

Despite the apparent success of the exorcisms and blessings conducted in Chamonix and elsewhere in the Alps, the glaciers began a series of retreats and advances that posed even greater dangers to valley inhabitants. Avalanches buried hamlets and herdsmen, along with their flocks, and floods were frequent and destructive. As time went by, however, most of the advances, when they came, were either equal to or not as great as the mid-seventeenth-century advances. Maps made in the early eighteenth century showed that all glacial tongues had withdrawn to such an extent that they no longer posed an immediate threat, although in 1820 and again in 1850 the Mer de Glace edged to within 50 meters (160 feet) of some occupied houses in Les Bois. Thereafter, however, it was apparent that a corner had been turned. The Mont Blanc glaciers would continue to advance and retreat, essentially scouring the land that had once been occupied, but it was always one step forward and two steps back. By 1873, the snout of the glacier had "hidden itself behind the rocks and been reduced to a blue strip, above the torrents between two cliffs worn smooth by ice."[4] By the early 1900s, only the tip of the Mer de Glace could be seen from the priory at Chamonix, looking more like a tail than a tongue.

In fact, it *was* the tail end of something, a roughly 350-year-long period between the fifteenth and mid-nineteenth centuries when average temperatures in the Northern Hemisphere fell an average of 0.6°C (1°F)—more at higher altitudes. Today this period is known as the Little Ice Age, so christened by Dutch-born glacial geologist François Matthes in 1939 in a short paper that appeared in the *Transactions of the American Geophysical Union.* Matthes originally invoked the term to refer to the dramatic advance of glaciers following a warm period called the Holocene Climate Optimal, which occurred between 9,000 and 5,000 years ago.[5] But somehow the term was extended to include the cooling period beginning in the late Middle Ages.

At any rate, Matthes clearly intended the Little Ice Age to be more about glaciers than climate, but over the years, the distinction became blurred. The term has too often been misconstrued to describe one long continuous cold spell, which it most definitely was not. Moreover, there is disagreement among scientists as to when the Little Ice Age began and when it ended, and even what caused it and why it quietly and quickly went away. No wonder, then, that scientists sometimes refer to the Little Ice Age as a period of "neoglaciation," meaning recent glaciation after the last Ice Age. Still, in addition to the record of glacial advances, enough evidence has been amassed from historical records, personal accounts, and the art and literature of the period to indicate the Little Ice Age was a real climate aberration in which temperatures were, overall, cooler than they are today.

Beginning around 1370 CE, Greenland was cut off from the world by impenetrable sea ice. By 1450 all of the Norse settlers who had not escaped the island either starved or froze to death. In the North Atlantic, treacherous icebergs and spreading ice sheets disrupted fisheries and shipping around Norway and Iceland. The Thames River froze over nine times in the 1600s and five times in the 1700s, during which hardy Londoners organized "frost fairs," complete with enormous tent cities set up on the ice-bound river. People could go ice bowling, ice skating, or ice wrestling, and warm themselves either with a sheep roasting, a round of intoxicating beverages, or both. During these same periods, the canals and rivers in the Netherlands froze over; even New York Harbor froze solid in 1780.

For anecdotal evidence, one need only page through Charles Dickens's perennial *A Christmas Carol* to find the London of 1843 in the grips of a "piercing, searching, biting cold" in which, on Christmas day, it began "snowing heavily." Similarly, Dickens describes the London weather in *Nicholas Nickleby* as being "intensely and bitterly cold" and "the wild country round, covered with snow." Today, Londoners experience an average of three days of snow each year, with usually not more than an inch of accumulation.

The Little Ice Age is also famously well represented in the work of Flemish painters Pieter Bruegel the Elder (1525–1569) and his son Pieter Bruegel the Younger (1564–1638). Both were known for their detailed depictions of harsh winter reality. To father and son, it must sometimes have seemed that the entire world was in a deep freeze. Bruegel the Elder cast the Roman census of Bethlehem during the birth of Jesus in a bleak winter setting, while his son set the adoration of the Magi in a snowstorm.

But again, we must take care not to infer that the Little Ice Age was marked by entrenched cold spells year after year. It was actually punctuated by periods of warmth as well as intense bouts of rain, sleet, and hail. These conditions produced a string of unfortunate consequences. The planting seasons across Europe were shortened to the point that crop failure was ensured. Diseases like malaria sprang up and spread in the chronically boggy, mosquito-infested land. Famines were a frequent occurrence, sparking bread riots in France and cattle raids in Scotland. In parts of central and southern Europe, the evil times inspired the hunt for witches, who were said to be responsible for hail and the destruction of harvests. In one region of Germany alone, at least 1,000 people suspected of being witches or practicing magic were burned to death over a period of forty years, beginning in 1580.[6]

The most conclusive evidence for the reality of the Little Ice Age, however, is the growth of glaciers, and not just a few dozen glaciers, either. In a paper published in *Science* in 2005, Johannes Oerlemans, a glaciologist with the Institute for Marine and Atmospheric Research at Utrecht University in the Netherlands, presented his construction of a "temperature history" using 169 glacier length records from all over the world, some dating back to the early eighteenth century.[7] His results show that glacial advances peaked in the early 1800s, followed by a dramatic drop-off in lengths after 1850. The concurrent advancement of the glaciers prior to 1850 stands in sharp contrast to the obvious retreat of the world's glaciers today. Historical records, eyewitness accounts, paintings, sketches, tourist photographs, and now satellite imagery show that since 1850 the world's glaciers have in general been falling back. Oerlemans points out that for the period between 1860 and 1900 (for which there are 36 records), only one glacier advance was recorded, and for the period between 1900 and 1980, out of 144 glaciers, all but 2 had retreated.*

Ironically, the cold may have imparted at least one benefit to the art world (other than the Bruegels' beautiful paintings of snowy and frozen landscapes). It has been hypothesized that the superb tone Antonio Stradivari achieved with his incomparable violins made in the early 1700s was attributable to his using the wood of spruce trees grown in the southern Italian Alps that had been subjected to years of lowered temperatures. Longer winters and cooler summers produce a slower-growing tree with wood that has more even growth and density, something essential for making high-quality soundboards.[8]

Some scientists caution that many of the arguments for a Little Ice Age are too anecdotal, unreliable, or biased to form the basis of a worldwide climatological event. The evidence, they assert, either seeks to blame the climate for various human crises (disease, famine, crop failure) during the Middle Ages or ignores other contributing factors. But empirical evidence, such as the phenomenal growth of the Alpine glaciers, as well as those documented in Scandinavia, Scotland, Germany, Iceland, Canada, and elsewhere throughout the Northern and Southern hemispheres during the sixteenth and seventeenth centuries, testifies to some sort of sustained trend in global cooling. Add to this paleoclimate records such as tree rings and ice cores, and the signal becomes even more salient. Although the cause of this big chill is still debated, most scientists agree that the primary culprit was most likely our own faithful star.

---

*Oerlemans's results have more sobering consequences for us today. His temperature reconstructions show that in the first half of the twentieth century, the average warming was 0.5°C (1°F), which closely agrees with other estimates of the average increase in temperature near the Earth's surface. More damning evidence of global warming.

The energy output of the Sun was once considered so unchanging that astronomers referred to the amount as the "solar constant." Then, beginning in 1978, satellites equipped with radiometers—which measure the rate of energy coming from the Sun over a given area—revealed what the eye could not: the solar constant is not so constant after all. The output varies, as measured so far, by about 0.1 percent above and below its "constant" value of 1,366 watts per square meter (W/m^2). By every definition of the term, the Sun is a variable star, but just barely, especially when you compare it to other variable stars in the Milky Way.

Astronomers have logged tens of thousands of variable stars in our Galaxy. Some vary in brightness by quite a lot, while others flicker only slightly. The causes of such fluctuations are many. For some, it's simply the onset of old age. After several billion years, the thermonuclear core powering an old star is so nearly spent that it can no longer produce a constant outflow of energy to support the star's great mass. But for a while, it can compensate. As its gaseous outer atmosphere bears down from its own gravity, the pressure squeezes upon the core until it ignites a thin shell of surrounding hydrogen. This new breath of ensuing radiant energy causes the star's envelope to balloon outward until it thins and cools, whereupon it falls back again and the cycle renews itself. As the hydrogen supply dwindles, the star contracts even more and the core begins to burn helium. These pulsations, which result in the characteristic brightness fluctuations, can go on for hundreds of thousands of years before the star finally breathes its last, gently exhaling its outer shell into space and leaving behind an Earth-size, but very dense, cinder called a white dwarf.

If it's not the effects of stellar old age that produces variability, it's a calamitous relationship in which one star is paired with a more massive companion. A typical example is a close binary system containing an aging sunlike star and a dense, white dwarf companion. As the aging star evolves and expands, as described above, the gravitational pull of the white dwarf siphons off gas in a process called "mass transfer." Matter streams from the normal star and accumulates in a swirling accretion disk surrounding the white dwarf. The disk itself may have an inherent instability brought on by an irregular flow or a break in the mass transfer or by tidal forces between the two stellar components; either way, these instabilities can trigger a sudden outburst in brightness should parcels of gas from the disk fall onto the surface of the white dwarf. Astronomers call these volatile objects cataclysmic variables. In extreme cases, the gas may accumulate on the surface of the white dwarf and eventually reach a thermonuclear "flash point" in which the hydrogen gas is rapidly burned to helium, leading to a nova explosion. After the eruption, the pair settles down and the whole process starts anew until, maybe 10,000 years later, it erupts again.

Some stars naturally pulsate to varying degrees, either because they're old or because they're still very young and have yet to reach the stage where nuclear reactions are triggered in the core. Whatever the cause, the periods of these manic fluctuations can take days or years; they can be regular or irregular or even completely isolated events. Given that most of these stars represent certain extremes in stellar evolution, we're probably safe in assuming that life as we know it would be hard-pressed to gain a foothold on planets orbiting stars such as these. For life to evolve from bacteria into a sentient form takes billions of years; therefore a nonvolatile, stable star is essential. Despite its 0.1 percent flutter, the Sun is just such a star.

The exact cause of the Sun's minor variability has yet to be worked out, but astronomers know that it is tied in with the comings and goings of sunspots. Over the course of eleven years, give or take a year, the Sun produces a peak number of the dark irregular spots, then returns to producing hardly any at all. It is from peak to minimum—which also corresponds to the rise and fall in the Sun's magnetic activity—that solar radiance changes by about 0.1 percent. It was once thought that the Sun's radiance would fall when splotched with numerous dark sunspots, but that is no longer the prevailing view. Like the upswing in fever that accompanies an outbreak of smallpox or measles, solar output actually increases the more sunspots there are. The reason: sunspots are not the quiescent regions astronomers once thought, but localized strong magnetic fields that inhibit hot gases from rising from lower levels of the Sun before reaching its visible surface, called the photosphere. Associated with the production of sunspots are bright magnetic features called faculae (from the Latin word *facula*, or "little torch"). When the Sun is observed in mostly red, or hydrogen-alpha, light, these appear as bright linear features that snake erratically along the photosphere. When large groups of sunspots are present, their numbers grow and they tend to group around the spots and slither more actively. These bright features, in part, explain why the solar output rises when the number of sunspots increases. The spots are simply tracers of the hotspots in the Sun's underlying surfaces.

The first tangible evidence that the Little Ice Age might be related to the Sun emerged when scientists began comparing old sunspot records with temperature and weather records. Some of the earliest-known naked-eye sunspot reports date back to 165 BCE. These, coupled with aurora observations, can be correlated with a more active Sun. Johannes Kepler was probably the first astronomer to have observed a sunspot using a camera obscura, in April and May 1607, though at the time, he mistook the dark blot to be a transit of Mercury, in which the planet's silhouetted disk is seen crossing in front of the Sun. (A Danish astronomer, Reiner Gemma Frisius—teacher of Gerard Mercator—actually observed a solar eclipse in 1544 from Louvain, Belgium, also using a camera obscura, but reported no sunspots. This is not too surprising since his observations were made near the end of what astronomers now recognize as a period of low sunspot activity.) Subsequent observations of sunspots were also made with pinhole instruments and camera obscuras, but not until the advent of the astronomical telescope in 1610 did European astronomers begin earnestly keeping sunspot records. The telescope allowed observers to safely project a magnified and more detailed image of the Sun onto a screen, on which sunspots could be sketched and plotted on a daily basis. Galileo was the first to do this, in June and July 1612.

Still, even as telescopic observers of the Sun were on the rise, the sunspot record taken during the beginning of the seventeenth century remains scanty at best. It is complete enough, however, to indicate that the Sun was beginning an upswing in sunspot numbers that would peak around 1614. Another smaller peak would occur around 1625, followed by a larger peak in 1639–1640. There should have been another outburst of spots leading up to 1650, but it never came. In fact, very few sunspots were observed (that is to say, "recorded") for another sixty-five years. This dearth, which occurred during the height of the Chamonix Valley's glacier siege, is known today as the Maunder minimum, after Edward Walter Maunder, an English solar astronomer who studied the phenomenon in the 1880s. Maunder had been alerted by reading two science papers written by German astronomer Gustav Spörer, who identified a similar dip in naked-eye sunspot numbers between 1400 and 1530. This period is known as the

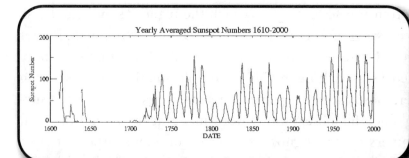

Figure 6. Sunspots come and go in cycles of 11 years. This chart shows the average number of sunspots between the years 1610 and 2000. The dearth in sunspots between the years 1645 and 1715 demarcates the Maunder minimum, a period that corresponds to the "Little Ice Age." Courtesy of NASA.

Spörer minimum. (An even earlier minimum, known as the Wolf minimum, is thought to have occurred between 1280 and 1350.)

With lower levels of solar output came fewer manifestations of solar activity, namely, aurorae, which are triggered when charged particles from the Sun—an *active* Sun—interact with atoms and molecules in the Earth's upper atmosphere, creating colorful displays of shimmering light. In the mid-1970s, a solar physicist named Jack A. Eddy confirmed Maunder's earlier report that there were little or no aurorae during the Maunder minimum. Checking old records, he found only 77 aurora sightings had been recorded worldwide when there should have been upward of 1,000 auroral nights in northern Europe alone.

There was still more evidence of a lethargic solar period, not written down in musty journals but etched in the growth rings of trees. Eddy's most profound contribution to solar-climate studies was synthesizing the work of previous scientists who had suggested a different but more direct link between solar variability and climate. As we saw with Stradivari's choice of wood for his violins' soundboards, long winters and cool summers yield slow-growing trees with narrower, denser growth rings than those in warmer, wetter climates. But not until 1961 did a Danish-American geologist named Minze Stuiver forge a climate connection between growth rings and the Sun.

It all centered around carbon-14, a radioactive isotope of normal carbon, and its journey from the sky to the Earth. Carbon-14 is produced in a chain reaction that begins in the upper atmosphere when high-energy cosmic rays from deep space ricochet off atomic nuclei, knocking off neutrons in the process. When the free neutrons meet with nitrogen-14 molecules in the lower atmosphere, they bond, creating carbon-14. The isotope eventually drifts to the surface, where it is absorbed by trees and plants and, in fact, all living creatures.

That carbon-14 could be found in living (and dead) organic matter was an accepted fact in the 1960s. Stuiver, however, was puzzled by the variations in the amount of carbon-14 he measured in ancient tree rings. Some rings had an overabundance of carbon-14, while others exhibited a scarcity. It was as if the precipitation of carbon-14, like rain and snow, came and went in spells. Eventually, he correlated the fluctuations to changes in solar activity. During periods of high solar activity, when the sunspot numbers are elevated, the solar magnetic field puffs out like a natural force field, shielding the inner solar system from the incoming cosmic rays. This, in turn, inhibits the production of carbon-14, as well as another isotope, beryllium-10, which is produced in the atmosphere when cosmic rays collide with oxygen and nitrogen molecules. (The beryllium-10 abundances have been measured in ice cores.) During quiescent solar periods, conditions are reversed. The Sun's magnetic field shrinks and the "shields" are down, allowing more cosmic rays to penetrate the Earth's atmosphere and thus enhance carbon-14 abundances in tree rings (and beryllium-10 in ice cores). Hence the presence, or absence, of these isotopes can be used as proxies to measure variations in solar activity in the distant past.

When Eddy measured the amount of carbon-14 in sequoias and bristlecone pine trees dating back to the Maunder minimum, he found unusually high abundances of the isotope, indicating a quiet Sun. This interlude in solar activity also corresponded to the extended period of lower-than-normal temperatures in the Little Ice Age. The Sun had left a telltale record of its rises and falls in activity within the living cells of trees and, in so doing, its effect on the climate. For Eddy, who, ironically, had set out to prove that the solar constant *was* constant, the discovery was a paradigm shift of the first order. In a 1976 article in *Science*, he wrote that the evidence "was one more defeat in our long and losing battle to keep the Sun perfect, or, if not perfect, constant, and if inconstant, regular."[9]

While the Sun may not be perfect, the data that supports this conclusion very nearly is. When sunspot numbers, aurora sightings, tree-ring growth, and isotopic abundances are compared with the climate record of the period, the fit is compelling. Prolonged dips in temperature are in step with prolonged dips in solar activity, which, in turn, correspond to higher abundances of carbon-14 and beryllium-10. The correlation also works in reverse, with lower levels of these isotopes matching elevated solar activity and extended periods of warmth. One of the most notable of these is the Medieval Warm Epoch (800–1300 CE). During its peak, between 1100 and 1250 CE, more aurorae and naked-eye sunspots were reported than in the three centuries before or after, and carbon-14 abundances were correspondingly lower. This multicentury stretch of mild weather proved to be both a boon to civilization at the time and a siren's call. Vintners in southern England were able to go head-to-head with French vintners by harvesting quality grapes from their own vineyards—something inconceivable today. The mild climate also beguiled the Vikings into settling colonies in Iceland and Greenland. Eventually, both the English vineyards and the Viking settlers would perish in the Little Ice Age, which followed on the heels of the Medieval Warm Epoch.

Given all this evidence, the question begging to be answered is, how could a solar variance factor of only 0.1 percent have played a role in the Little Ice Age or the Medieval Warm Epoch? It seems too much of a lightweight to be a real climate contender on its own. Most of the time, that is a true statement. But we're not talking about an eleven-year flicker of solar inconstancy; rather, we're talking about what might happen if irradiance gets stuck at some extreme range of output for an extended period, say, a couple of centuries. Only then can it begin to compete with the bigger forces of climate. As such, it plays the climatological wild card.

Without having the solar readings from an orbiting radiometer dating back to the Little Ice Age or earlier, scientists can only extrapolate back in time to estimate how much fainter the Sun was then. But who knows? Instead of being 0.1 percent fainter, it could have been at times 0.15 or 0.20 percent fainter. If such a decline persisted over many decades, it would be like moving the Earth that much farther away from the Sun for hundreds of years. Imagine if the Earth were suddenly relocated 30 million or 60 million kilometers (18 million or 37 million miles) farther from the Sun. The atmosphere would effectively hold in warmth for a time, but not for centuries. The cold would eventually win out.

Our wild card might also swing the other way, and here prospects could be equally grim. The solar irradiance value does not just translate into how bright the Sun appears to the eye. What is measured as 0.1 percent output at visible wavelengths is actually some 10 to 20 times greater at ultraviolet wavelengths. That much radiation, if sustained, would erode the Earth's protective ozone layer, allowing more ultraviolet radiation to penetrate the middle levels of the atmosphere and reach the surface. One can only imagine how a sustained increase of solar output at visible and ultraviolet wavelengths might affect our climate, not to mention life on Earth. This would not be good news any time, but it would be particularly bad news if the upswing were superimposed onto the current trend of global warming.

There's that term again. As it is strictly an anthropogenic (caused by people) problem, not a cosmogenic one, global warming technically does not fall within the purview of this book. On the other hand, I am forced to raise it because some scientists claim that it is the Sun, not greenhouse gases, that has contributed to the observed 0.6°C (1°F) rise in temperature throughout the twentieth century. But precise irradiance measurements of solar output only go back a few decades, and extrapolating back more than a century is fraught with mathematical uncertainty. So where is the evidence, if any?

Those who believe the Sun has little to no effect on global warming (I'll call them the no-Sun camp) make up the vast majority of the world's climate scientists. Their case is straightforward enough. True, the chilling of the Little Ice Age was likely the result of solar forcing, or the lack of it, over a period of several centuries, but the sudden and rapid warming of the twentieth century, particularly since 1980, can be best explained by anthropogenic forcing—that is, human-induced effects such as burning fossil fuels and large-scale tropical deforestation. The slow and inexorable rise of greenhouse gases such as carbon dioxide far, far outweighs any component due to changes in solar output.

Those in the pro-Sun camp, of course, have data sets that they claim say otherwise. For example, an oft-cited 2003 study found that the Sun has brightened steadily during the past two decades by an amount that would account for half or more of the given warm-up of the twentieth century.[10] And in a more recent study, another group of researchers used these results as a starting point for their own research, which applied statistical methods to analyze how Earth's atmosphere responds to slightly stronger solar output. Although these researchers concede that greenhouse gases may be partly responsible for global warming, they argue that the Sun may have contributed to the problem by as much as 30 percent.[11]

But there may be a flaw in their data. They base their claim on measurements from radiometers flown aboard various satellites. Although these instruments are painstakingly calibrated on Earth, not all radiometers are alike. Suffice it to say that they vary by such factors as wavelength and power range, sensor diameter and resolution. There are scanning and nonscanning radiometers, some with wide fields of view, others employing detector arrays. Given these various characteristics, it should come as no surprise that they have differing scales of sensitivity. But on top of this, the radiometers themselves can change while in space, due primarily to the degradation of their optical elements and detectors. This necessitates recalibrating the detectors on orbit—*if* they can be recalibrated—and typically without regard to radiometers aboard other satellites. These disparities introduce what is called "instrument drift" when you compare the datasets of different radiometers. To be able to splice together these datasets in order to produce a long-term record of solar irradiance, instrumental drift must be taken into consideration—something, critics say, the brighter-Sun studies did not do.

How much more confounding can this issue become? It turns out that there actually is evidence to suggest that the Sun has not only increased in radiance since 1940 but has sustained it. A group of scientists have been able to reconstruct sunspot records as far back as 11,400 years ago by dating radiocarbon concentrations in tree rings, and 7,000 years ago based on variations in the strength of Earth's magnetic field (which is affected by variations in solar activity), as recorded in ancient magnetic minerals such as magnetite. What they find is that the level of solar activity during the past 70 years is "exceptional."[12] The previous period of equally high activity occurred about 8,000 years ago. Moreover, almost all of these highly active periods were shorter than the present phase. During the past 11,000 years, the Sun was similarly active for only about 10 percent of the time; for the 7,000-year period, it rivaled today's activity between 2 and 3 percent of the time.

So does this mean that the Sun *is* causing or compounding global warming? Neither, says S.K. Solanki of the Max Planck Institute for Solar System Research and lead author of the *Nature* paper announcing this discovery. He and his colleagues point out that, despite this exceptionally high activity, it is unlikely that the Sun's activity has contributed to global warming since 1970. In fact, based on the durations of previous episodes, Solanki thinks there is only an 8 percent chance that the current elevated level will extend for another 50 years and a 1 percent chance of it lasting until the end of the century.

Other scientists have taken an altogether different approach to understanding how solar irradiance might affect the climate, one that is unrelated to global warming. The Sun, after all, is a star. Why not measure the activity of other sunlike stars? Long-term monitoring of these stars during periods of low activity might provide insights into low-activity cycles of the Sun. It sounds like a long shot, but that remains to be seen. The problem is that, to date, no stars have been found that are *exactly* like the Sun in all key respects. It's not that the Sun is unique—it's just that finding a star that could pass as the Sun's twin, point for point, is not all that likely. Such a star would almost have to have been born at the same time and out of the same cosmic dust cloud as the Sun, to have emerged from the same stellar gene pool, so to speak; it would also have to be a single star and it would have to have a temperature, size, and luminosity similar to the Sun's. In the 1980s there were a number of stars considered to be good candidates, but they were identified before accurate distances were obtained with the Hipparcos satellite in the early 1990s. Incorrect distances make for overestimating or underestimating a star's luminosity, diameter, and age. Basically, without knowing a star's distance precisely, it is a cinch that you will end up with an erroneous picture of its evolutionary status. And that's what happened. By 2004, most of these stars were found to be far too old to be used as solar analogs.

However, the Hipparcos data did show that there are several dozen near matches—including several, in fact, now known to sport Jupiter-size planets. Still, when one comes closest to matching the Sun's spectral type (i.e., color) and luminosity, the candidate doesn't mesh with its mass or chemical content. Probably the two stars that can be considered the Sun's nearest twins are 18 Scorpii, in the northern realms of the constellation Scorpius, and HD 98618,* located in the cup of the Big Dipper. Both have chemical profiles similar to the Sun, but both are also about 2 percent more massive, hotter by 0.7 and 1.1 percent respectively,

---

*"HD" stands for the Henry Draper catalog of stars, published between 1918 and 1924.

more luminous by 3 and 6 percent respectively, and are about half a billion years younger than the Sun.[13] It is not clear what would happen to us if we could substitute the Sun with a star that is slightly hotter, larger, and more luminous, but it is likely that it would have a deleterious effect on the Earth's ecosystem. (On the other hand, stars such as these would be excellent candidates for harboring terrestrial planets and, perhaps, some form of life.)

Unfortunately, none of these otherworldly insights offers much in the way of predicting what might happen in the years to come on Earth, particularly with global warming wielding the most influence. Furthermore, other forces besides greenhouse gases exert their effects on climate in ways that remain mysterious and unknown. Theories abound. Some scientists claim aerosols from ground-based pollution and jet aircraft are creating clouds like giant mirrors that reflect sunlight back into space, in effect dimming the amount of sunlight that reaches Earth. Others lay the blame on cosmic rays. Nevertheless, the vast majority of scientists agree that when compared to the anthropogenic effects of global warming in *this* century, people, not the Sun, have tripped the climate alarms. I suppose that about all we humans can say in our defense is that it was not our intention to denature Earth's climate, but that's what happened. Maybe future generations, more magnanimous than we, will understand and forgive us. Maybe not, and who could blame them?

Whatever effect the Sun may have in the coming climate drama, one thing is certain: Earth will continue on its natural course regardless of what we do or what it does or what happens to it. The geologic record tells us that the Earth has made and remade itself countless times, long before we came along to scratch on cave walls or intentionally fly passenger jets into buildings filled with innocent people. The human race may abide on this planet for another million years, hugging close to its religions, its arts, its sciences, and whatever else defines and shapes it, but Earth does not require it. Neither do the stars. Look anywhere beyond our little nook of the Galaxy and you will see a universe that is not only dispassionate, but dangerous and random. Comets plow into planets. Stars explode without regard to what clinging forms of life may be in the vicinity. Black holes suck up space and time at will. That we're here now is a wonderful accident, but one of these days we may find ourselves in the way of the greater cause of Nature. It may come from the deepest reaches of outer space, or we may find that it lies as close as our own Sun.

CHAPTER 3

A TEMPERAMENTAL SUN

A [geomagnetic disturbance] caused by a large high-altitude nuclear det-
onation is similar in many ways to that of solar-induced geomagnetic
storms except that a nuclear-caused disturbance would be much more
intense with a far shorter duration.
— Oak Ridge National Laboratory, "Electric Utility Industry
Experience with Geomagnetic Disturbances," 1991

On August 2, 1972, the Sun dramatically revealed a volatile, and distinctly disconcerting, character trait. A series of powerful solar flares went off like flashbulbs on the Sun's surface starting that day and over the following ten days, sending a steady stream of highly charged particles Earthward. These outbursts occurred at a time when Apollo missions were being regularly flown to and from the Moon; *Apollo 16* had returned to the Earth in April, and the *Apollo 17* crew was gearing up for what was to be the program's final mission in December. Fortunately, no astronauts were on the Moon's surface at the time. As powerful as some of the flares were, particularly the August 4 flare, it is estimated that in a very short time the astronauts' organs could have absorbed as much as 400 rems (roentgen equivalent man), even inside the thinly shielded lunar module. A dose of 400 rems within a few hours' time would probably have been enough to make them wish they'd stayed in bed; in fact, it might have killed them. A diagnostic chest X-ray exposes patients to an average of 0.010 rem or less; the accepted occupational dose limit for radiation operators in the health field is no more than 5 rems—per year.

Looking back, it's amazing that during the entire Apollo era no major solar-particle events occurred to threaten the astronauts. The time spent on the lunar surface was brief, to be sure, but considering that the Sun peaked in activity around the time of the *Apollo 11* landing and was still going strong through the end of the program in 1972, their good luck might just as easily have turned bad. As long as they were confined to the heavily shielded service module, they remained fairly insulated from solar particles; at worst, they might have absorbed a total of 35 rems in their bones and spleens. On the Moon, however, either in the lunar module or ambling over the surface, they were potential sitting ducks. Such lessons in hindsight are why more formidable types of shielding than that used for the lunar modules are being considered for future missions to the Moon and Mars. It's a dangerous solar system.

But that's to be expected in space, particularly interplanetary space. One would expect to encounter the worst kind of solar radiation out there, well away from the shielding effects of the Earth's atmosphere and magnetic field. What's the worst that can happen on the ground? Maybe a power outage here or a communications glitch there. Maybe a little jitter on the Internet, but nothing more serious. Right?

Most of the time, yes. But there have been some notable exceptions. In 1989, a strong solar flare initiated a blast of high-energy particles that struck Earth on March 13 at 8:20 p.m. eastern standard time, setting off high-voltage alarms, overheating transformers, and tripping capacitors throughout the northeastern part of the United States and Canada. At the Hydro-Québec Power Company in Montreal, the intense electrical convulsion overwhelmed and shut down seven static compensators (which stabilize voltage flow) in one minute and overloaded transformers, prompting the tripping or deactivation of reactive power compensators. By 2:45 a.m., the system collapsed in a power blackout, leaving more than 6 million people without electricity. Had the event occurred during the higher-load winter or summer peaks, the outages might well have spread down the East Coast of the United States. As it was, it took more than nine hours to restore 83 percent of the system.[1] An April 2002 threat analysis by Canada's Office of Critical Infrastructure Protection and Emergency Preparedness estimated that approximately 19,400 megawatts of power in Québec and millions of dollars in revenue had been lost due to the outage. The report stated, "[Solar] storms of less severity occurred in September 1989, March 1991, and October 1991, but they were still strong enough to hinder utility operations. These events have caused industry to become more aware of the reality of geomagnetic storms and the destruction they can wreak on unprepared infrastructure components."[2]

Though these kinds of geomagnetic disturbances are relatively rare and astronomers' monitoring of the flares that cause them has increased, the world's communications and power infrastructure remains extremely vulnerable. Since the second half of the twentieth century, power transmission lines in North America and throughout the world have become more interconnected and greater in length. On the one hand, this interconnectedness helps safeguard power grids from localized failures. But should a solar storm take down one or two systems, those connected to them may fail as well. Even continental oil pipelines, like the Alaskan pipeline, are susceptible to geomagnetically induced currents, creating false readings in flow meters and even physical erosion of the pipeline itself.

Events like the March 1989 storm dramatically demonstrate how the Sun's connection to Earth's geomagnetic environment—or magnetosphere—can have unwanted consequences on the ground. Most of the time, the constant "wind" of particles blowing outward from the Sun at supersonic speeds flows around the magnetosphere and trails off downstream, creating a "magneto-tail" that extends out into interplanetary space. An explosive flare, however, releases a flurry of high-energy particles that can get trapped in a kind of death spiral around the field lines of the magnetosphere. As these whipped-up particles interact with atmospheric atoms and molecules, they emit photons, producing vivid displays of the northern or southern lights.

As astronomical phenomena go, solar flares are fairly common. They might not be as common as sunspots, but like sunspots their increased frequency indicates an active Sun. Each eleven-year cycle of solar activity produces at least 20,000 solar flares of varying strength, 100 of which may temporarily disturb Earth's magnetosphere, creating a geomagnetic storm. Judging by these numbers, it doesn't appear that solar flares by themselves pose any real problems on the Earth's surface, and, generally, they do not. But the flares that are especially bright in X-rays are almost always sema-phores of another geomagnetic drama, which goes by the clinical name of "coronal mass ejection," or CME. These can be as disruptive as the March 1989 storm—more so under the right conditions.

Simply put, a coronal mass ejection is a piece of the Sun ejected into space. The concept of mass being ejected by the Sun has been theorized for over a century and is thought to have been captured in a drawing made by Italian astronomer Gugliemo Tempel during the total eclipse of 1860 over Spain.[3]

The first definitive observation of a CME was recorded in December 1971 by a coronagraph aboard NASA's Orbiting Solar Observatory 7. A coronagraph is a telescope that uses a disk to block out the Sun and produce an artificial solar eclipse— allowing the Sun's rarefied atmosphere, called the corona, to be

Figure 7. One of the first depictions of a coronal mass ejection, drawn by Gugliemo Tempel during the total eclipse of 1860 over Spain. The ejection is the round feature at lower right. Available at http://sunearthday .nasa.gov/2006/locations/firstcme.php.

clearly observed. The exact physical process leading up to a CME is still not well understood, but most scientists agree that changes in the Sun's magnetic fields trigger the sudden release of mass and energy. On average, a CME vents 100 million tons of plasma into space at speeds of between 400 and 2,700 kilometers (250 and 1,600 miles) per second. Depending on its velocity, the leading edge of the plasma shell can reach Earth in a matter of hours or days, resulting in a light show in the form of an aurora and a geomagnetic storm of some intensity.

The relationship between solar flares and coronal mass ejections has, in recent years, become more confusing and not a little controversial. Conventional wisdom argues that the kinetic energy released by a CME is far greater than its associated flare, but astronomers want to know if this is, in fact, true because, as is often argued, a solar flare is not even needed to trigger a CME. On the other hand, there appears to be an almost one-to-one relationship between the occurrence of very powerful X-ray flares and big, bad CMEs. Is that merely circumstantial evidence? A new generation of solar satellites, such as the NASA-sponsored Solar Radiation and Climate Experiment (SORCE) and the Reuven Ramaty High Energy Solar Spectroscopic Imager (RHESSI), are now able to accurately quantify the amount of kinetic energy released by both the flare and its associated CME. SORCE made the first such measurement during a huge flare on October 28, 2003: the estimated total radiated energy was 4.6×10^{32} ergs.[4]

By itself, an erg is a very small unit of energy. To power a 100-watt lightbulb for an hour takes about 3 trillion ergs (3×10^{12}, or 3 followed by 12 zeroes).[5] But put them together exponentially, and ergs soon become less trivial. For example, 1 ton of TNT releases an average value of about 4×10^{19} ergs of energy. The Hiroshima bomb had a yield of about 13,000 tons of TNT.

Hence, Hiroshima's erg value is $13,000 \times 4 \times 10^{19} = 5.2 \times 10^{23}$ ergs. So the amount of total radiated energy cited for the October 28, 2003, flare was around 900 million times more than the Hiroshima bomb, give or take. It's also worth noting that the CME arising from that event was simultaneously observed by a NASA/European satellite, the Solar Heliospheric Observatory (SOHO), and was determined to have released 1.2×10^{33} ergs.[6] This value is a little over 2 billion times more powerful than the Hiroshima yield.

At the very least, astronomers now have a lower limit on energy released by a flare. Still, some fifty satellite observations later, the conclusions continue to support the notion that, at least for larger events with fast CMEs and bright X-ray flares, the total energies produced by both are comparable. Additionally, astronomers observing flares and CMEs near the Sun's limb have found that most of the flares appear to form somewhere beneath the erupting filament that becomes the heart of the CME, with the degree of centrality higher for the brightest X-class flares, a finding that is consistent with many flare-CME models.[7]

Even with these findings, however, the process leading up to a solar flare and a CME remains elusive. Indeed, the spatial placement of flares and CMEs remains in question because the only ones that have been observed are those on or near the Sun's limb, leaving the true three-dimensional arrangement in doubt. About the only thing one can say with any certainty is that, along with sunspots, both events arise as by-products of the Sun's magnetism. We can only make our apologies to the theorists and try to explain the basics of these phenomena here.

The breeding ground for a flare is a group of sunspots. On the simplest level, when sunspots appear in pairs, one spot generally has a positive magnetic "footing" (or pole) and the other has a negative footing. Put a lot of sunspots together, as often happens when solar activity is high, and you have a complex tangle of horseshoe-shaped magnetic field lines called coronal loops. A plasma of charged particles is accelerated through the loop from one magnetic footing to its opposite. Solar astronomers call this the flare's precursor stage.

One might also call it the precipitate stage because it fore-shadows the coronal mass ejection to come. The Sun's surface is an extremely complicated place, constantly roiled by rising and falling currents of gas beneath the surface. These currents distort the shape of the sunspots' magnetic field lines so that their foot-ings constantly shift. The greatest amount of distortion takes place when a new sunspot group forms among a preexisting group or two groups collide, forming into rows of loops that look like a toy Slinky—its rings pinched together on one side. The crowding squeezes the innermost magnetic "footings" into smaller and smaller regions until they are nearly touching. At this point, the pinched poles create an induction current of gas, which in turn generates enormous heat throughout the loop, between 10 million and 100 million degrees C (18 million to 180 million degrees F). Fountains of hot gas dance wildly beneath the expanding coronal arch, some forming miniature loops. At this stage, clumps of very hot plasma may be seen racing up into the loop as bright momen-tary streamers. Within an hour or so, the energy is such that the loop begins emitting X-ray and gamma-ray radiation, usually from near its crown, but sometimes along its sides.

What happens next happens relatively quickly (although exactly *how* it happens is still under considerable debate). Stressed beyond their limits, the magnetic lines abruptly snap and then reconnect, forming a new coronal loop. This process unleashes enormous amounts of stored magnetic energy and, like releasing an overinflated balloon, propels a blob of trapped plasma through

Figure 8. Coronal loops of multimillion-degree gas extend 300,000 kilometers (186,000 miles) above the surface of the Sun. These structures are often precursors of potentially disruptive geomagnetic events in Earth's atmosphere. Courtesy of NASA and the Transition Region and Coronal Explorer (TRACE) team.

the corona and out into space. This is the mass-ejection event. At the same time, material falling back to the surface of the Sun may unleash similar energies near the Sun's surface, creating the flare. This is the big finale of the impulsive stage. The coda, the decay stage, is simply the diminishing of the flare's intensity, as the new magnetic field relaxes into a less distorted shape.

Whether or not a flare and a CME will affect Earth depends, among other things, on the amount of kinetic energy that is released, where the event occurs with reference to the Sun-Earth line, and the orientation of the CME's magnetic field to Earth's. If the CME occurs on the limb, or apparent edge, of the Sun, it manifests itself as a spectacular plume of filamentary gas, two to three times larger than the solar disk itself, but the impetus is not in Earth's direction; it's going to miss us by hundreds of millions of kilometers. If, however, the flare and the subsequent mass ejection happen to be pointing near or toward Earth—and especially if the orientation of the CME's magnetic field has a charge that is opposite that of Earth's—the result is a one-two punch beginning with another type of solar storm called a solar proton event, or SPE, followed by the arrival of the CME. As the name implies, the particles in an SPE consist largely of high-energy protons and electrons. They are accelerated in the CME shock while still close to the Sun, no more than a few solar radii out, and race along the "tracks" of the interplanetary magnetic field at near-light speeds. Generally, they can reach Earth in less than an hour, or sooner in some rare cases. Their unpredictability and rapid onset make them a decided threat to satellites and astronauts. Two such SPEs occurred on October 28 (as mentioned above) and November 4, 2003. The October 28 event, called the Halloween Storm, "snowed" SOHO's coronagraph with particles, temporarily blinding the satellite's camera. The November 4 flare that preceded the SPE was also, to date, the most powerful flare ever recorded. Astronauts on board the International Space Station were forced to retreat five times into the better-shielded aft part of the station to escape radiation.[8]

The intensity of solar flares can be measured in a number of ways, but when it comes to the kind of flares that produce the most energy in X-rays—which is the kind that is the most worrisome—the GOES* satellite series is the great quantifier. Since 1976, different detectors aboard various GOES spacecraft have

---

*Geostationary Operational Environmental Satellites circle the Earth in a geosynchronous orbit, which allows them to remain in a fixed position at an altitude of 35,800 kilometers (22,245 miles) to monitor large areas of Earth's surface.

been measuring and tracking these energetic events. Technically speaking, these kinds of X-ray solar flares are classified according to their X-ray brightness in watts per square meter within a wavelength range between 0.1 and 0.8 nanometers, or billions of a meter. The least-powerful X-ray-class flares are classified as B or C flares and have little consequence on Earth. The particles from medium-strength M-class flares can cause temporary power blackouts in the polar regions and intermittent radio communications on the sunlit side. But it is the powerful X-class flares that are the ones to watch out for.

Since 1978, only 29 solar flares ranked X9 and greater have been recorded, with 1991 having a record 8 such flares. Within the X class of flares, these are extremely powerful events. Still, of late, there seems to be a trend upward in intensity, which may be due to better monitoring or, perhaps, something else intrinsic to the Sun. In any case, on March 6, 1989, a flare (not associated with the March 13, 1989, geomagnetic storm) estimated to be a class X15 overwhelmed one of the GOES X-ray monitors. That record stood for four months, until an X20 flare erupted on August 16 of that year. A flare event on April 2, 2001, was also classified an X20. The November 4, 2003, flare broke all official records, coming in as a class X28.[9] Because the GOES detectors were completely saturated at the flare's peak, some scientists think that the flare was even stronger, perhaps between X40 and X45.[10]

No other flares brighter than this have ever been measured, though one is suspected. On September 1, 1859, British astronomers Richard Carrington and R. Hodgson independently observed a starlike eruption within a cluster of sunspots. In his report to the Royal Astronomical Society, Carrington described its brilliancy as being "fully equal to that of direct sunlight," and in the same report Hodgson said it was "much brighter than the

Sun's surface." Both men also noted that magnetometers at the Kew Gardens Observatory in London registered disturbances minutes after the event. Carrington, however, remained skeptical that the flare and the magnetic storm were related. "One swallow," he wrote, "does not make a summer."[11]

The flare itself lasted all of 5 minutes, but the total amount of energy released by the 1859 flare is estimated to have been 10^{32} ergs. This is some 2 billion times the energy of a 1-megaton hydrogen bomb (and still less than one-tenth the energy emitted by the Sun each second). Moreover, the subsequent CME reached Earth in an astonishing 17 hours and 40 minutes, which means that it was pretty much directed fully at Earth and magnetically well connected. The auroral displays lasted several nights and could be seen as far south as Cuba and Hawaii in the Northern Hemisphere and as far north as Santiago, Chile, in the Southern Hemisphere. As the storm's magnetic field intensified, it induced electric currents in telegraph wires. At times the current was so great that telegraph operators could communicate with each other cross-country without even using their batteries. In some cases, the currents overloaded the wires, setting off a spate of fires in the United States and Europe.[12]

Naturally, Earth's upper atmosphere took the biggest hit, something scientists can now quantify with other flare events. It works like this: A pulse of increased solar radiation creates a brief enhancement in atmospheric nitrates via chain reactions between solar particles and molecules of nitric oxide. These, in turn, are captured by aerosols that fall to Earth, where they deposit themselves in thin but distinct layers on polar ices. By correlating nitrate concentrations measured in Greenland ice cores with more recent flare events, scientists can derive a pretty good estimate of the energy released in those flares. The more energetic the flare, the greater the nitrate deposits. Guided by this, scientists estimate that the energy released in the 1859 flare was 6.5 times greater than that released by the March 1989 flare (the one that doused the lights along the northeastern coast of the United States and Canada) and that it destroyed 3.5 times more ozone.[13] That last detail has far-reaching implications: erode the ozone layer, which resides tens of kilometers high in the stratosphere, and you erode the very shield that prevents harmful ultraviolet radiation from space from reaching Earth's surface and your skin.

Scientists also estimate that the 1859 flare was nearly as energetic as, and perhaps equal to, the great August 1972 event, which could have spelled disaster to Moon-roving Apollo astronauts. And it could very well happen again.[14] More sobering still, our time span of such energetic flare events is quite limited (spanning only a couple of centuries at best); hence it's possible that more powerful events have escaped our notice.

About right now a fair question to ask is, could the Sun produce flares even more powerful than the 1859 one—and hence correspondingly more destructive coronal mass ejections? Bruce Tsurutani, a plasma physicist at NASA's Jet Propulsion Laboratory who has studied the 1859 flare event, thinks it unlikely, although he hedges a bit. He has heard speculation from some colleagues about "super flares," which could potentially spell big trouble for Earth's space weather. "At 10^{37} ergs, some people have indicated that there would be a mass extinction due to the destruction of the ozone layer by the X-rays," he says. "However, there is no evidence for an event of this type happening in any of the records that scientists have looked at. The Sun and the magnetosphere are of a finite size and have finite magnetic field strengths and therefore cutoff energies. At most, my guess is that a solar flare *might* be as intense as 10^{35} ergs."[15]

And yet astronomers know of at least nine solar-type stars (similar in color, mass, rotation rate, and age to our Sun) whose own flares make the Sun's biggest flares seem like party poppers by comparison. They can be 10 to 100 million times more energetic, last for up to a week, and cause the star to brighten up to 1,000 times its normal luminosity. If a similar flare suddenly occurred on the Sun, the inner planets (including our own) would be reduced to cinders, and the ice on Jupiter's and Saturn's moons would melt to shallow seas before freezing again.

Why would other sunlike stars flare up so violently while our Sun does not? One theory is that these super flares are caused by the disconnection and reconnection of magnetic field lines of two stars in a close orbit. There is plenty of evidence for this. A class of stars called RS Canum Venaticorum stars (RS CVn being the prototype star in the constellation Canes Venatici) are very nearly sunlike in their properties, except they all possess close-by stellar companions with orbital periods of between 1 and 14 days, and they all emit flares far more powerful than the biggest flares of the Sun. In December 2005, astronomers using the Swift satellite observed a powerful eruption on the RS CVn star cataloged as II Pegasi. The flare was so bright in X-rays, scientists at first thought it was a gamma-ray burst (one of the most powerful stellar explosions known). It triggered a false alarm on Swift's Burst Alert Telescope and saturated its X-ray Telescope. II Pegasi's companion has a mass less than half that of the Sun and lies but a few stellar radii away.[16] The tangle of the stars' magnetic field lines no doubt creates dramatic flare events, but tidal forces from their mutual gravity also spin up their rotation rate to once in 7 days (versus 28 days for the Sun), which enhances their flare power.

Of the nine known sunlike stars with super-bright flares, however, none appear to have very close stellar-size companions.

Their rotation rates, too, are similar to the Sun's leisurely 28-day period. According to two astronomers who have studied these stars, Eric Rubenstein and Bradley Schaefer, one possible explanation is that the flares are spawned by a close-in planet the size of Jupiter.[17] A companion of more modest mass would not spin up the primary star, nor pull on it as much as a stellar-size object would. All the companion need do is anchor the magnetic field, which it could achieve if its own magnetic field were of sufficient strength. Just like the RS CVn flare stars, those with close-in planetary companions could experience super-flare activity.

In a way, our planet supports this hypothesis. The Earth has apparently never suffered a flare of at least 10^{36} ergs in the last two billion years because such an event would likely show up in the geologic record as a mass extinction. Of the five great mass extinctions known, none can be attributed to such an event, although some scientists argue that the Ordovician mass extinction that took place 443 million years ago might have been caused by a gamma-ray burst (more about that later). Being primarily a radiation event, a sudden flare-induced mass extinction would manifest itself in the rock sediments as changes in the layers of fossilized plant and animal life, but absent the metals and dust from an asteroid impact or the ash from a volcanic explosion. So here's another reason why our Sun may be as quiet and stable as it is: it never acquired a companion near enough and massive enough to trigger super flares.

It is an interesting exercise to imagine how different life on Earth might be if the Sun had a companion star. It certainly is a could-have-been scenario, since some two-thirds of the stars in our Galaxy possess companions, and, of the 300-plus known extrasolar planets, a dozen or so reside in binary systems. Moreover, most sunlike stars have companions. Fortunately, we seem to be an exception. The solar system wouldn't be such a life-friendly place if every century or millennium or so a companion induced the Sun to let loose with enough energy to sterilize the innermost planets. On the other hand, perhaps the Sun *does* have a companion, a possibility some astronomers occasionally wonder about. If its mass were, say, a tenth of the Sun's and it never ventured much closer than a light-year (nearly 6 trillion miles) every 30 million years or so, we might not even know it existed.

More comforting is the behavior of a star called 18 Scorpii, which is barely visible to the naked eye in the constellation Scorpius. This star is considered by some to be one of the most promising candidates for harboring a life-bearing planet.[18] Like the Sun, it has no stellar companion and it rotates slowly, once every 23 days. It is a little hotter than the Sun (and hence may be younger) and has a 7-year, rather than an 11-year, activity cycle, during which 18 Scorpii's brightness varies by 0.09 percent (compared to the Sun's 0.10 percent decadal variation). If more of these stars could be observed over decades of time, astronomers would be able to reveal a range in the various properties and luminosities of sunlike stars that, in turn, would tell us more about the nature of our own home star.

The good news is that, as cosmic threats go, we're not likely to be irradiated out of existence (at least by the Sun). The bad news, though, is that we remain vulnerable to strong flares and CMEs. Strong flares have occurred in the not-too-distant past, and there's no reason to think they won't happen again. Any way you look at it, a perfect solar storm as powerful or more powerful than Carrington's 1859 flare could be catastrophic today. You can bet that major power grids and communications services would be disrupted. Given our dependence on electronics, computers, cell phones, and GPS devices, we could end up deaf, dumb, and blind to anyone not in our immediate vicinity. Ultraviolet and X-ray radiation from the outburst would heat up and expand the atmosphere on Earth's day side, exposing more neutral atoms to solar ionization, and in essence create a denser "electron content" in the ionosphere. These alone wouldn't hurt us physically, but low-orbit satellites would experience increased drag, which could result in the loss of computer tracking capability and even the satellite itself. And that's not including potential damages to electronic switches, optical sensors, and guidance systems. At its worst, such a disaster could cost the telecommunications and electrical power industry several hundred billion dollars in damage and wreak unprecedented social havoc. Imagine having to deal with protracted power outages and not being able to communicate with other countries, other cities, or even family, friends, or colleagues across town for days or weeks. What might happen to social order under these circumstances?

In these troubled times, this is not a scenario anyone wants to dollop onto their plate of woe. After all, such an eventuality might not present itself within the foreseeable future. Still, the possibility of this occurring in our beleaguered generation, or that of our children, cannot be ruled out. This prompted me to check in with the United States Department of Homeland Security to see if anyone there was considering a contingency plan for such an event. Phone calls were made, e-mails were sent. My only response was an e-mail reply from someone who said my request for information had been referred to, of all things, the Office of Multimedia. This office, the reply stated officially, serves as the liaison to movies, television programming, books, videos, and other multimedia-type projects. "I have checked into your specific request and have not been able to find [a] DHS Component that has the requested information."[19] Perhaps they thought I was scripting a movie or writing a science fiction novel? I took another stab at it by doing some serious Google mining. That effort eventually led me to a 2004 Department of Homeland Security Science and Technology directorate document addressing plans to protect US infrastructures from diabolical terrorist threats.[20] Under "New and Emerging Threats and Vulnerabilities," one potential technological menace caught my eye: "Electromagnetic, directed energy and pulse weapons which use no ammunition and are unrecognizable by most law enforcement personnel."

The authors of the report, of course, are not referring to the sudden jolt of electromagnetic radiation from the airburst of a nuclear bomb that would, in turn, overload electronic systems and bring down civilization as we know it. The logistics of launching bombs across the United States to their target makes them undesirable, even to wild-eyed terrorists. The DHS report was likely referring to the development of some sort of portable weapon capable of creating widespread electromagnetic pulses that could potentially disrupt the communications infrastructure. (A shoulder-launched electromagnetic pulse bomb?)

Of course the Internet is a gold mine of information on how such electromagnetic bombs might be constructed and delivered, but, truth be told, their reality remains largely in the realm of science fiction. Insofar as government officials associate them with clever terrorists, such scenarios probably should not be pursued to any great extent... *except*, perhaps, when it comes to coronal mass ejections from Carrington-like solar flares. Now we're talking about events that could really trigger indiscriminate electromagnetic pulses and, within a few hours, cause widespread chaos. Personally, I think such a hypothetical scenario qualifies as a prepare-for-the-worst-approach, similar to the approach that should have been taken long before Hurricane Katrina struck the Gulf Coast in August 2005. We were obsessed with looking for terrorists then, too. Instead, we were blindsided by Nature.

We've seen that the Sun, like Earth's climate, is complex, unpredictable, and ever changing. Our Sun is a *star*, a fairly stable one, true, but certainly not the "constant" we thought it was, because, in fact, it never was. Nowhere is this more evident than in the frequency and potency of its outbursts. The timing of solar cycles has always been predictable, but not so the occurrence of major solar flares and CMEs. This has been particularly true over the last two solar cycles, in which the outbursts have been unexpected and unexpectedly energetic.[21] What's happening? Is this merely a blip in behavior or a trend? As yet, no one can say, and that's what makes the future Sun a roll of the dice.

There are certain life stages that astronomers know the Sun will cross, but what remains in doubt is when those stages will occur, how they will overlap, and how they will affect life on Earth. The drama will play out over the next 5 billion years, as the Sun ages, destabilizes, and dies. But consider that long before the death of the Sun, life on Earth, in whatever form, will have reached its endgame. In fact, the processes leading up to this eventuality are already under way. Astronomers have long known that, over time, as the Sun continues converting hydrogen to helium, its core will become denser, hotter, and more luminous. This process isn't noticeable now, but given that the Sun has already reached middle age, over the course of the next billion years it will become all too obvious. Every 100 million years or so from now, the Sun will become brighter by 1 or 2 percent, maybe more later. As solar luminosity increases, so too will the Earth's surface temperature. The rise in temperature will increase evaporation, raising the atmosphere's water vapor content. This, in turn, will trap outgoing infrared radiation, creating a greenhouse environment that will make the world hotter and wetter still.

What happens next depends on how you like to factor things like plants and people into the mix. One scenario, proposed in 1992, predicts that the increased rain will weather rocks at a faster rate, dissolving carbon dioxide and locking it into calcium carbonate sediments on the sea floor. Eventually, perhaps as soon as 500 million years from now, carbon dioxide levels in the atmosphere will be too low to support megafaunal life (that's you and me). Photosynthesis will slow and finally cease altogether. But if carbon dioxide starvation doesn't get us, the heat will. According to this same model, global mean surface temperatures could increase from 50°C to 100°C (90°F to 180°F) in less than 200 million years, which would pretty much shut down whatever was happening with the carbon cycle at the time.[22] Meanwhile, another scenario suggests that our very existence could keep things habitable for a little longer. It predicts that carbon dioxide produced by Earth's plant life will maintain the planet's habitability for at least 800 million years, maybe more, after which the carbon dioxide levels begin to fall and catastrophic overheating has its way with the biosphere.[23] The coup de grâce occurs as global surface temperatures approach 100°C in about 1.2 billion years. That's when Earth's oceans evaporate, volcanic carbon dioxide accumulates in the atmosphere, and Earth's climate becomes more like that of hellish Venus. And you thought global warming was bad!

Still, half a billion years, give or take, is a good long time, long enough for humans to solve their societal shortcomings and anthropogenic screw-ups (like global warming) and move on to bigger and better things. It's an intriguing exercise to contemplate how much humankind might achieve if it can survive over a geologically long time period. All that has gone on before in the world, what we call history, would be nothing more than a prelude to such an extended future. In the next chapter, however, we will consider a cosmic connection that could potentially foreclose on that future. We wouldn't have to wait a million years for it to occur, either, or even a century, for that matter. In fact, it could happen at any time.

CHAPTER 4

AT ANY TIME

As a natural disaster, and a very infrequent one, the impact of comets and asteroids is less likely to evoke concern than ozone depletion or global warming, because the latter are looked upon as man-made disasters, and the likely time scale for those problems to become serious is far shorter. This ignores one important fact about the impact hazard, that it is random and could occur at any time.
  —P. R. Weissman, "The Comet and Asteroid
    Impact Hazard in Perspective,"
    from *Hazards Due to Comets and Asteroids,*
    T. Gehrels, editor

Let's go burn down the observatory so this will never happen again!
  —Bartender Moe Szyslak in *The Simpsons* episode
    "Bart's Comet," in which Bart discovers a comet
    that is about to hit Springfield.
    This is Moe's solution to the crisis.

Friday night, October 10, 1992, was an exceptionally fine night for high school and college football in the northeastern

United States. The autumnal sky was cool and clear, and a beaming, full Hunter's Moon floated low in the eastern sky. Games were being played that evening from Anne Arundel County, Maryland, to Olmstead Falls, Ohio; from Pittsburgh, Pennsylvania, to Falconer, New York. As dusk waned to dark, most of the fans were riveted to the action on the field, but at 7:48 p.m., a few in the stands were suddenly drawn to something they saw moving out of the corner of their eyes. Looking up, they were stunned to see a fiery tail of debris as bright as the Moon arcing steadily out of the southwest. Some witnesses were alerted to the object's flight when they heard the rumble of sonic booms. For nearly half a minute the thing flared, flashed, and fragmented along its flight path, giving at least 17 amateur and professional videographers time to record a few seconds of the event. Finally, only a handful of fading tracers were left continuing along the long, sloping trajectory toward the northeastern horizon. The unearthly sight left many witnesses speechless, but not all. "What the hell was that?" a Fairfax, Virginia, witness exclaimed through nervous laughter. "Beats the hell out of me!" replied his rattled companion.[1]

Meanwhile, eighteen-year-old Michelle Knapp was sitting on the sofa in her family's home in Peekskill, New York, watching television with her boyfriend when a thunderous explosion shook the ground and rattled the windows. At first, the startled couple thought something had struck the house. They tentatively ventured outside, not knowing what to expect, and soon discovered that the right side of the trunk of Michelle's red 1980 Chevy Malibu was crumpled in like a soda can; smoke was coming from somewhere. Shattered plastic and glass, wiring, and bits of car trim were scattered all about the ground. The housings for the brake and back-up lights had been blown out and lay dangling beneath the car. Just below and behind the bumper, in a shallow,

Figure 9. The Peekskill bolide as it appeared over Altoona, Pennsylvania, on October 10, 1992. Photo taken by S. Eichmiller of the *Altoona Mirror*, image courtesy of the *Altoona Mirror*.

slightly elongated crater, was a dark brownish gray rock about the size of a football. One side of the rock was smeared with the car's red paint. When Michelle touched it, she said it felt warm.

The police were called and then, at the telltale odor of gasoline fumes, a fire truck was summoned. Not knowing what else to do, the police filed a report of criminal mischief, and the offending rock was impounded. The next day authorities released it to scientists at the American Museum of Natural History in New York, who determined that the object was a 12.4-kilogram (27-pound) "ordinary" chondrite, or stony meteorite. Later, analyses by nine research teams of the meteorite's history—how long it had been exposed in space to radionuclides, noble gases, and cosmic rays—showed its age to be about 27 million years. Before it hit the Earth's atmosphere, it had been approximately 200 meters (650 feet) across and weighed some 10,000 kilograms (22,000 pounds).[2] Based on thousands of eyewitness reports and video recordings from various angles, the Peekskill meteorite was one of perhaps as many as four or more fragments that reached the ground that day, although none of the others struck any property.[3] Indeed, no others have ever been found.

The Peekskill event was highly unusual. People see shooting stars, or meteors, all the time. In fact, several times each year, if you place yourself in the right place at the right time, you can witness a display of meteors called a meteor shower, which occurs when Earth passes through the gritty debris path left in the wake of a periodic comet. Some of the more popular meteor showers are the Perseids in August, the Leonids in November, and the Geminids in December (the genitive case referring to the constellation from which the meteors appear to radiate—in these cases, the constellations Perseus, Leo, and Gemini). Occasionally, these showers produce meteors that flash and fragment and leave behind brief smoke trails.

Less frequent are the lone, fiery, longer-lasting meteor events called fireballs or bolides, which the Peekskill meteor clearly was. Another famous one was seen August 10, 1972, in broad daylight, streaking south to north over the western United States and Canada. The brilliant meteor was visible for over a minute and a half before it skipped off the top of Earth's atmosphere and back into space, leaving a fading smoke train in its wake.

When a meteor does make it to the ground, at which point it is called a meteorite, it is possible to physically recover and relate it back to its blazing trail in the sky. According to meteorite databases such as the Meteoritical Bulletin, the Natural History Museum's Catalogue of Meteorites, and MetBase, well over 1,000 documented meteorite falls since the ninth century have been associated with meteor sightings. A boulder-size object falls to Earth once a week on average, and yet recorded instances of meteorite falls that injure people or damage property, like Michelle Knapp's car, are extraordinarily rare. In fact, the only authenticated incident of a meteorite striking a person occurred in Sylacauga, Alabama, on November 30, 1954. Thirty-four-year-old Ann Hodges was taking an afternoon nap on her living room sofa when she was awakened by an explosion. An eight-pound stony meteorite penetrated the roof and ceiling of her house, bounced off a large radio console, and struck her on the left hip and arm. Hodges suffered a serious grapefruit-size hematoma on her hip, from which she recovered (although she was said to suffer emotional scars from the event for the rest of her life).*

Reported deaths and injuries associated with meteorite falls exist but are apocryphal at best. There's the tale of a seventeenth-century Franciscan friar who died from injuries caused by a meteorite striking his thigh; a dubious report of a Kentucky farmer being killed by a meteorite while sleeping in his bedroom the night of January 14, 1879; and the report that an eight-pound meteorite struck the ship *Malacca* sailing from Holland to Dutch Batavia (now Jakarta, capital of Indonesia) in 1648, killing two sailors. There's also an account of a dog being killed by a meteorite during a shower of stones in the village of Nakhla, near Alexandria, Egypt, in 1911. Buildings, houses, mailboxes, cars, and cows have all reportedly been conked by meteorites with varying effects. But when it comes to alleged meteorite deaths, the Chinese have them all beat. Records from 616 BCE to 1915 CE contain seven accounts of meteorite fatalities, some in which a rain of stones wiped out whole towns![4] True though some of these reports may be, none can be confirmed or refuted.

Given the number of verified close calls that are on record, death or injury by meteorite cannot be considered totally inconceivable, but the odds are clearly against it. The Peekskill fall, in

---

*Ironically, across the street from the Hodges home was the Comet Drive-In Theater, which was prominently decorated with a huge neon sign showing a comet streaking toward the heavens.

which no one died, was considered by itself to be a one-in-a-billion event. On the other hand, the idea that meteorites can cause serious mischief has recently been given serious thought. In 1997, the National Transportation Safety Board sponsored a study in response to the mysterious crash of TWA Flight 800 off the US East Coast on July 17, 1996, to determine whether a meteorite could conceivably bring down an aircraft. The conclusion was that a hull-penetrating meteorite strike to an aircraft over the United States might happen once in 59,000 to 77,000 years.[5] Humanity has been flying for only a little over a century, so I suppose it's too soon to start blaming meteorites for aviation mishaps.

What about meteorite mishaps to spacecraft or even astronauts? That's a different story. In space, micrometeoroids—small particles no heavier than a gram—can move between 5 and 68 kilometers (3 and 42 miles) per second. At that velocity, even a grain of sand might carry some lethal punch, depending on where it strikes. Scientists have used portions of spacecraft returned to Earth, like part of the Hubble Space Telescope's solar panel array and the insulation blankets of the EURECA space-environment satellite, to analyze these "hypervelocity" impacts. They have found that most are caused by space debris—that's humanmade junk like fragments of metal or plastic from explosions, paint flakes, dust, lost tools, and, yes, astronaut waste. Fewer than a tenth of the impacts are attributable to natural micrometeoroids.[6] Nevertheless, both space debris and micrometeorites pose potential threats to space exploration and are keeping engineers busy designing new ways to protect spacecraft and space suits from them.

For example, during the August 2007 STS-118 shuttle mission, a micrometeoroid that was probably an orbital debris particle completely penetrated one of the shuttle *Endeavour's* aft radiator panels, leaving deposits from the impact, but fortunately no damage, on the payload bay door. Although such impacts are not considered unusual, the damage of this particle was larger than any previously seen on the shuttle's radiator panels, which are 12.7 millimeters (0.5 inch) thick. The entry hole was measured at 7.4 mm by 5.3 mm (0.29 by 0.20 inches), and the particle that did the damage was estimated to have been 1.55 mm to 2 mm (0.06 to 0.08 inches) across, your typical grain of beach sand.[7] As of late 2008, the US Space Surveillance Network, charged with keeping track of space debris, reports a total of 12,851 pieces of space junk from whole payloads to centimeter-size fragments.[8]

Our own self-made space debris can occasionally pose a real threat to people on the ground, too. Notable cases have been the reentry of Skylab in July 11, 1979, most of which fell into the Indian Ocean but which also scattered debris over Western Australia; the Solar Maximum Mission satellite on December 2, 1989, which also fell into the Indian Ocean but without littering any continent with space junk; the spectacular February 7, 1991, reentry of the Salyut 7 space station that lit up the skies over Argentina but also rained debris on the town of Capitán Bermúdez, 280 kilometers (175 miles) northwest of Buenos Aires; and the Compton Gamma Ray Observatory, which reentered and fell without incident into the Pacific Ocean on June 4, 2000.

More serious was the uncontrolled reentry of a Soviet spy satellite Cosmos 954 on September 18, 1977. Fortunately, the debris fell upon a sparsely populated region in the Northwest Territories of the Canadian Arctic. It could have been much worse. To power its ocean-scanning radar used to track US ships at sea, Cosmos 954 was carrying its own compact nuclear reactor with 110 pounds of enriched uranium-235, enough to be lethal to anyone who came within a few hundred feet, had it spilled out onto the ground. At the time, American scientists were reported to have commented that if the satellite had failed one pass later in its orbit, it would have struck near New York City at the height of rush hour.[9] Shades of Peekskill! Although many of the larger pieces of the satellite were eventually recovered, the thawing of the ice in the spring suspended further search operations. There may still be fragments out there glowing in the Canadian wilderness.[10]

In February 2008 another space junk drama unfolded, albeit one with military overtones. The US Department of Defense shot down a disabled, out-of-control spy satellite that it said was due to crash uncontrollably. Officials cited its fuel tank full of solid hydrazine, a toxic chemical, as needing to be ruptured in order to prevent anyone on the ground from getting hurt. The missile shot was successful, and the satellite apparently disintegrated safely into pieces before reentering the atmosphere. But this is the kind of thing that may become all too commonplace in future years. Satellites malfunction or run out of fuel, orbits decay, gravity wins. There's a *lot* of stuff up there that will one day come down again.

But it's not just intact satellites and abandoned space stations that pose a danger; it's also rocket stages, fuel tanks, and other discarded pieces of spacecraft jetsam that fall back to Earth. Most of it breaks up at altitudes of around 78 kilometers (48 miles) and burns up in the atmosphere, especially stuff made of material with a low melting point like aluminum. But larger, denser components, or objects contained within some sort of housing, break up at lower altitudes and if made of material with a higher melting point—say, titanium, stainless steel, and beryllium—may reach a terminal velocity before burning up, which means they will fall to Earth.

It seems our ventures into space have created an unintended artificial cosmic connection, one that clearly is not beneficial to ourselves or the planet. It may be that one day some sort of space debris–management system—an orbiting garbage truck or an attachable robotic maneuvering device to facilitate a controlled reentry—will be developed to clean up after ourselves. Until then, if the solution is to shoot down satellites before they enter Earth's atmosphere, that may end up doing more harm than good because in blowing up something in space, even in low-Earth orbit, you create a flurry of additional debris particles moving at ballistic velocities, all with new trajectories. When the Chinese government used one of its discarded satellites, Fengyun-1C, as target practice for an antisatellite system in January 2007, it created one of the worst satellite breakups in the history of the space program, generating more than 2,600 fragments larger than 10 centimeters (4 inches). Estimates suggest that there could be as many as 150,000 particles as small as 5 millimeters (0.2 inch), exceeding all predictions.[11] All these pieces may be in orbit for years yet to come.

If you want to quantify the probabilities of death or injury by meteorites or even space debris, you have your work cut out for you. You must look at both the rate at which lethal stones fall to Earth and how the world's population is distributed around the globe. Although the latter can be fairly well determined, the former is a bit more challenging. To name just a few factors, the lethal rate will depend on the mass and type of meteorite, its trajectory and velocity, direct versus indirect strikes on human flesh, and even, say some scientists, the seasons. It's been estimated that between 1,000 and 40,000 tons of meteoric material enter the Earth's atmosphere each day, most of which is either vaporized or falls to Earth as dust. But good luck trying to quantify how many stone-size objects reach the ground. Estimates range wildly, but a 1984 study by a group of Canadian astronomers probably quantifies it best. They combed through nine years of archived photographic observations from a network of all-sky cameras and concluded that eight events per year drop at least 1 kilogram (2.2 pounds) of meteorites in an area of a million square kilometers (386,000 square miles).[12] This same group also predicts that at least sixteen buildings are damaged by meteorites every twenty years and that one person will be struck every nine years.[13] Based on these numbers, it's safe to say that, except for the occasional freakish close call, the chances of being killed or injured by a meteorite are slim to none. You're much more apt to be dolloped with bird poo.

There is, however, another factor to consider, one that would certainly increase the lethality of a meteorite: its size. Imagine how different Peekskill would look today if the stony meteorite that fell there weren't the size of a football but something more on the order of a football field. *Now* you're talking real death and destruction, for at least a 5-mile radius from ground zero. And if that same meteorite were made of iron, like the one that created the iconic Meteor Crater in Arizona, you could write off life and property for at least 1,000 miles from the impact site.

We're at a point in this chapter where we have to make a fuzzy distinction. Before meteorites were meteorites they were first *meteoroids*, rocks moving in orbit around the Sun. Meteoroids only become meteorites when they smack into the surface of a planet. Depending on its composition, trajectory, and velocity, a meteoroid 1 meter (3.2 feet) in diameter in space may end up the size of a football if it survives its passage through the Earth's atmosphere. But as meteoroids get bigger in size, from boulders to knobs, massifs, and mountains, you can't call them meteoroids anymore. At some point, you have to call them what they are: asteroids.*

How likely is an asteroid impact on Earth? Once again, the answer is: not very. On the other hand, we're no longer consid-

---

*Technically, the world's largest-known meteorite is the Hoba meteorite, discovered near Namibia, South Africa, in 1920. Its dimensions are 2.7 meters (8.8 feet) long by 2.7 meters (8.8 feet) wide and 0.9 meters (3 feet) deep. Its mass is estimated at 60 tons. It still lies where it was found and is preserved as a national monument.

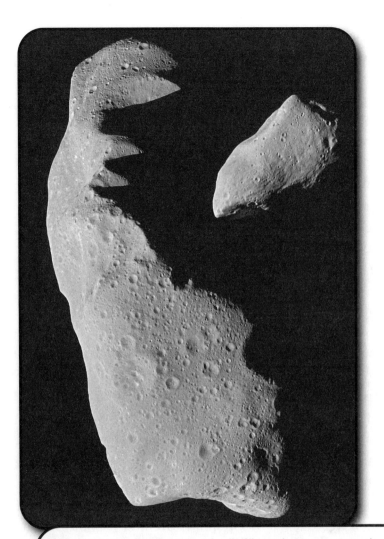

Figure 10. The asteroids Ida (left) and Gaspra are shown to scale in this composite image taken by the *Galileo* spacecraft while en route to Jupiter. Ida is 58 kilometers (36 miles) long, and Gaspra about 17 kilometers (10 miles). These asteroids are just two of the billions of such rocky objects that orbit the Sun, mainly between Mars and Jupiter. Impacts with objects as large as Gaspra and Ida are capable of causing mass extinctions on Earth. Courtesy of NASA/JPL-Caltech.

ering the probability of just one person being felled by a cosmic potato, but tens of thousands, even millions, of people getting killed by the impact of a small mountain *nearly simultaneously.* What's even more chilling is that something that big would likely be detected by astronomers months in advance and, given our present technology, we couldn't do anything about it except watch it approach and enter the atmosphere to do its worst. To be proactive about diverting or destroying an incoming asteroid, we'd need years to prepare, and even then, there's no guarantee. This has perennially been the stuff of great, and not-so-great, cinema.

But it has basis in reality. Most asteroids orbit the Sun in a well-defined "asteroid belt" between the orbits of Mars and Jupiter, but others are on erratic paths that can take them sweeping right across the orbit of the Earth. For instance, in late 2004 astronomers discovered that an asteroid some 330 meters (1,000 feet) across was going to come close—very close—to the Earth. At first, they gave it a scary 1-in-300 chance of hitting our planet on April 13, 2029, which also happened to be Friday the thirteenth. But then images taken of the asteroid years before were examined, and its old position with respect to the fixed starry background could be compared with its new position, allowing astronomers to refine their calculations about where it would go. This time the news was better. The latest numbers show that asteroid, designated 99942 Apophis, will miss Earth by about 30,000 kilometers (23,000 miles), or one-tenth the distance to the Moon, a clear miss. Still, Apophis will pass near enough to be seen moving across the sky in one night.[14] Apophis, in fact, returns to the Earth's neighborhood in 2036 but will likely miss us that time by over 49 million kilometers.[15] Nevertheless, this asteroid remains at the top of the watch list.*

---

*The name *Apophis* is the Greek derivation of the Egyptian god Apep the Destroyer, patron of evil, darkness, and chaos.

Fortunately, Earth's cosmic shores have been relatively free of huge, threatening asteroids or comets for the past million years or so, and, so far as we know, none, other than Apophis, perhaps, pose any future danger. From time to time, however, we are reminded that we should never be lulled into a false sense of security. On October 5, 2008, a small asteroid thought to be between 1 and 5 meters (3 to 15 feet) in diameter was discovered and tracked in space by astronomers around the world until, two days later, it entered Earth's atmosphere over northeastern Africa. Objects this size enter the atmosphere on average every few months, but this was the first time an asteroid's plunge toward Earth had been predicted. Although nothing could have been done to prevent its colliding with Earth—even if it had been 100 meters or more across—a prediction, at the very least, would have bought us a little time to make some preparations in the strike zone.

In the future, scientists hope such predictions will become more commonplace, because although we are less likely to be caught off guard by the approach of a large asteroid, a smaller but potentially dangerous object—a city smasher—might slip in under the radar. This has, in fact, happened several times, although all of them, fortunately, were close calls. In one instance, an asteroid with the potential kinetic energy of between 1,000 and 2,500 megatons of TNT was detected *after* it had missed the Earth by less than 700,000 kilometers (435,000 miles). Other near misses by such objects occurred in 2002 (88,000 kilometers, or 55,000 miles); 2003 (465,000 kilometers, or 290,000 miles); and 2008 (537,520 kilometers, or 334,000 miles). One of the nearest misses on record occurred in March 2004 when a rock 30 meters (100 feet) across brushed by at a distance of 43,000 kilometers (26,500 miles), which is about one-ninth the mean distance between the Earth and the Moon. Such near misses are thought to occur once every 2 to 3 years.[16]

Sometimes, however, maybe once every 1,000 years, Earth gets blindsided by a small, sneaky object that one might describe as a glorified meteorite. Because these objects cannot be detected until they're practically upon us, or after they impact, they pose a real hazard. And though the damage they do is more localized, it can be nonetheless considerable.

The most recent example of such an event occurred about 7:15 a.m. on June 30, 1908, when an object of some kind exploded over a largely uninhabited region of Siberia near the Stony Tunguska River. The fireball could be seen in a clear sky from a distance of 160 kilometers (100 miles) with a trajectory inclination of about 15°. Upon impact, the thunder was deafening, even from a distance of nearly 500 kilometers (300 miles). The shock wave from the explosion had an estimated kinetic yield of between 3 and 10 megatons of TNT, with recent research favoring the lower range of these values.[17] Whatever the impactor's true yield, the blast undeniably flattened more than 2,000 square kilometers (770 square miles) of Siberian forest, broke windows hundreds of kilometers away, and rattled seismic detectors across Eurasia. Witnesses were knocked senseless to the ground, storage huts were incinerated, and a group of reindeer herders still sleeping in their cedar bark tepees about 32 kilometers (20 miles) from the impact were blown into the air, tepees and all. A herd of 700 reindeer near the explosion's epicenter were burned to ashes. Only a few deaths have ever been attributed to the explosion, thanks to the remote location, but it's a wonder many in the vicinity weren't killed.[18]

The "Tunguska event," as astronomers call it today, was undoubtedly a cosmic connection of exceptional proportions, but its exact nature remains one of science's biggest mysteries. Much of what is known about the fireball and its behavior at impact comes from eyewitness accounts, some of which have been passed down to their descendants. The indigenous people who inhabited the region at the time, the Tungus, believed in shamanism, a practice that might be described as a form of interacting with and summoning the power of spirits, both good and bad. The Tunguska event, then, was interpreted in spiritual terms not in their favor. They believed the explosion was retribution brought upon them by a powerful shaman in a rival clan, something that was no doubt underscored by the eerie display of glowing night skies for many weeks thereafter. For decades, they believed the region where the impact occurred was bewitched, and few were foolish enough to venture there.

Despite their mystical belief systems, the observations of the Tungus people were invaluable in describing a rare natural event that scientists, unfortunately, were much too late in investigating. The first scientific expedition to the epicenter wasn't made until 1927, nineteen years after the explosion. The results only deepened the mystery. The expedition leader, Leonid A. Kulik, who virtually founded meteoritic science in Russia, was shocked to see a stark landscape of felled trees lying in a radial pattern as far as the eye could see. The trees that remained standing looked like bare sticks. Kulik was intent on locating the meteorite, which he assumed lay embedded in the swamp at the epicenter, but neither he nor follow-up expeditions ever found any traces of metal that could be attributed to a meteorite. The lack of an impactor led scientists to conclude that the object exploded before hitting the ground; but this, too, compounded the mystery.

No fragment of the projectile has ever been found, and its nature, either a small comet or an asteroid, is still debated. The argument for its being a comet, made primarily of ice, is based largely on the mechanism of its disintegration and aftermath. Most scientists agree that the Tunguska object exploded at an altitude of no more than 10 kilometers (6 miles). But those in favor of the impactor being a comet counter that a large meteoroid (or small asteroid) would have broken up much sooner during its passage through the atmosphere, at an altitude of at least 100 kilometers (62 miles). A rocky object of between 10 and 50 meters (30 and 160 feet) would have flattened and its edges flared backward during its fiery passage through the atmosphere. Subsequent shock waves would have broken the object into numerous pieces, a sight that would have resembled the Peekskill fireball but on a much larger scale. For a time, the fragments would have plummeted together, bound by a common aerial shock wave, but upon approaching terminal velocity each would have veered from its pre-atmospheric trajectory, as determined by mass and aerodynamic lag, striking the ground in a shower of stones.[19] Such an event occurred on February 12, 1947, once again in Russia, but this time in the Sikhote-Alin Mountains near the village of Paseka, which is some 440 kilometers (270 miles) northeast of Vladivostok. This, too, was an intensely bright daylight fall accompanied by a deafening explosion. But in its aftermath, inhabitants found meteorites in an elliptical-strewn field of 1.3 square kilometers (0.5 square miles), the largest forming a crater 26 meters (85 feet) across by 6 meters (20 feet) deep. Known as the Sikhote-Alin fall, it ranks as the largest meteorite shower in history and the twentieth century's second-most powerful cosmic impact.

One can appreciate, then, why the absence of a meteorite in the Tunguska event supports the idea that the impactor was a comet. Like the Sikhote-Alin fall, the breakup of a meteoroid should have littered the region with numerous stony splinters and stones, with the largest fragment producing a crater.[20] But aside from what some claim to be tiny meteoritical particles embedded in trees, no meteorites have ever been found. As a comet body, it would have had an initial mass of some 2 million tons, an entry velocity of 31 kilometers (19 miles) per second (twice the velocity usually cited for an asteroid), and, upon exploding, a kinetic energy yield of about 15 megatons of TNT (comparable to the estimated asteroid yield). The shock wave from the airburst would have been more than sufficient to flatten miles of forest in a radial pattern, while the icy, dusty cometary body would vaporize, leaving no trace.

The argument that the Tunguska event was the result of an asteroidal body is largely predicated on the relation between its entry trajectory and, extrapolating backward, its original orbit around the Sun. (This same approach, however, has also been used to support the cometary origin.) A fireball's path through the sky is, in a sense, like a tracer that leads back to its source or, in this case, its orbit. Given enough visual information from observers in different locations about its path through the sky, its inclination to the horizon, and the background stars, the orbital elements of a meteor can be calculated. For example, based on the excellent videographic record and eyewitness accounts of the Peekskill meteor, astronomers were able to determine that it came out of the constellation Libra. Its orbital path was inclined only about 5° to the plane of the ecliptic (the mean plane of the Earth's orbit around the Sun), and it rounded the Sun once every 1.8 years. Its degree of eccentricity, a measure of how "out-of-round" its orbit is, was 0.41. These parameters are normal for asteroidal bodies. Most known comets, however, have orbits that are generally more eccentric (0.5 to 0.9) as well as more inclined to the plane of the ecliptic (5° to 25° or more). Short-period comets have orbital periods from a few to around 200 years. Long-period and undiscovered comets have periods longer than 200 years and even millions of years. These objects are not gravitationally enslaved to the ecliptic, and many exit the inner solar system never to return. A newly discovered comet can seemingly come out of nowhere at any time. In sum, from all the eyewitness accounts of the direction and inclination of the Tunguska fireball, astronomers have been able to model hundreds of possible orbital elements. They find that the fireball's trajectory mirrors the less eccentric, less inclined, asteroidal paths.[21]

Without a meteorite, the asteroidal explanation for the Tunguska event seems untenable. If one could be found, though, it would settle the matter, or at least strengthen the argument, which is why scientists keep looking. In 2007, a team of Italian scientists proposed that Lake Cheko, which lies about 8 kilometers (5 miles) north-northwest of the Tunguska epicenter and on a line extending from the trajectory entry of the bolide, might be a lake-filled crater. For one thing, the shape of the lake's floor bears a striking resemblance to that of the Odessa meteor crater in Texas. Seen in cross section, the floor of the Odessa crater is funnel-shaped, with its maximum depth near the center. The floor of Lake Cheko is similarly funnel-shaped, unlike many of the Siberian lakes, which are formed by melting ground ice in the soil above the permafrost, creating water-filled depressions. Typically these lakes are shallow with a steep slope and a flat floor.

Another intriguing feature of Lake Cheko is that seismic analysis indicates the presence of a "prominent reflector" 10 meters (32 feet) below the lake floor at the center. Could this be a fragment of the projectile? Crater scientists remain skeptical, citing the lack of shocked rocks and evidence of melting, common features of true impact sites. Moreover, accounts handed down to descendents from eyewitnesses hold that the lake was known before the impact, also throwing doubt on the claim. But what is one to make of the fact that Lake Cheko doesn't appear on maps of the region prior to 1929? Should that be chalked up to maps that lack finer detail? It looks like the matter will be settled only by extracting a core sample, which the researchers aim to do.[22]

It should come as no surprise to learn that there are also more exotic explanations for the Tunguska event: an encounter with a miniature black hole, for one, or the annihilation of a small bit of antimatter in the Earth's atmosphere. All have been ruled out, though the theories continue to fly. Most scientists, whether they subscribe to either a comet or an asteroid as the impactor, view Tunguska as a sobering reminder of what could happen suddenly and without warning if even a small asteroid struck the Earth. How different, they ask themselves, would the modern perspective be had this object struck Moscow or New York City? Given the amount of kinetic energy it released, the death toll would likely have been in the tens of thousands. Even a small asteroid, between 10 and 30 meters (32 and 100 feet) wide, has the potential to release the equivalent of 3 megatons of TNT if not vaporized during its plunge through the atmosphere. Obviously, the impact of a modest-size asteroid in any populated region would be incredibly destructive and would make an effective, though tragic, display of the dangers of asteroids.

Scientists repeatedly warn that a collision with an asteroid is inevitable. They can say this because they have studied the crater record on Earth and the Moon and know something of the rate of impacts. A Tunguska-like event is thought to happen once every 1,000 years or less (thus there is a 1-in-10 chance of a similar Tunguska event happening this century);* a projectile capable of forming the iconic 1.86-kilometer (1.1-mile) diameter Barringer

---

*Although if you lower the object's size and the amount of energy released, the frequency increases to once every 100 years.

crater in Arizona may come every 20,000 years. Objects capable of producing hazardous tsunamis—200 meters (650 feet) across—strike the Earth every 100,000 years. The really big planet-busters, at least 10 kilometers (6 miles) in diameter, are thought to collide with Earth every 200 million to 300 million years, leaving behind a crater up to 200 kilometers (120 miles) in diameter. The last such impact occurred around 65 million years ago off the coast of Mexico's Yucatán Peninsula. That impact, and its aftereffects, are thought to have wiped out the dinosaurs.[23]

The problem with these numbers, however, is that though they are large and therefore imply that the occurrence of asteroid impacts on Earth is rare, you nonetheless cannot with any accuracy predict when an impact will occur inside the given frequency range. Just because a Tunguska explosion is calculated to happen once every 1,000 years doesn't mean it couldn't happen tomorrow or 200 years from tomorrow. The reason is simple and somewhat disconcerting: astronomers simply don't *know* how many Tunguska-like Earth-crossing asteroids in the 100-meter to 1-kilometer range remain to be detected. Fortunately, concerted searches use telescopes around the world that sweep the skies nightly for this celestial flak. As of 2004, about 2,700 Near Earth Objects (NEOs) had been cataloged. To date, more than 5,500 have been detected.

Before going any further, we need to sort out some necessary, but sometimes confusing, nomenclature relating to the class of potentially hazardous interplanetary bodies. Some of the following terms may seem interchangeable, but they're not. The big umbrella term already mentioned, NEOs, covers both asteroids and comets with perihelia (minimum distances from the Sun) of 1.3 astronomical units (AU), or 1.3 times the mean Sun-Earth distance. In round numbers, 1.3 AU is 195 million kilometers, or 121 million miles. Subsumed within the class of NEOs are Potentially Hazardous Objects, or PHOs. These are asteroids and comets that pass within 0.05 AU (less than 7.5 million kilometers, or 4.6 million miles) of the Earth's orbit and have diameters greater than 50 meters (160 feet). Asteroids, not including comets, that are NEOs are collectively referred to as Near-Earth Asteroids or NEAs. Near-Earth Comets are, that's right, NECs.

Estimates vary, but on average astronomers think there may be more than 20,000 NEAs with sizes in the 140-meter (460-foot) range that remain to be discovered.[24] In a very real sense, being more prevalent than the class of "civilization killers" makes them more dangerous because, as mentioned, an object of this size could appear on our horizon with little or no warning. Moreover, obtaining an impact rate of these objects on Earth is problematical because their impact signatures can be difficult to identify. Out of the 174 or so known impact craters on Earth,* only 19 are less than 1 kilometer (0.6 miles) across. The dearth of smaller impact sites is most likely due to erosion or, as in the case of the Odessa crater in Texas, the fact that it was used as a target range, a dumping ground, and a source of road fill before it was elevated to being a natural landmark and a tourist attraction.

Oddly enough, your chances of being killed by a rare planet-destroying asteroid are better than you might think. Clark Chapman, a planetary astronomer with the Southwest Research Institute in Boulder, Colorado, and David Morrison, with NASA's Ames Research Center in California, in 1995 devised a way to compare the chances of being killed by asteroids to other causes of death. In their initial findings, they discovered that people were

---

*The number of impact craters on Earth is variously cited as being between 140 and 190, depending on the source. The number given here was obtained from the Earth Impact Database maintained by the Planetary and Space Science Centre, University of New Brunswick, Canada. The database may be accessed at: http://www.unb.ca/passc/ImpactDatabase/.

more likely to die by car accident, homicide, and fire (in that order). But they also found that the chances of being killed by a global impactor were greater than being killed by a terrorist act, an insect bite, a natural tsunami, or an earthquake.[25] Chapman has recently updated his calculations, which now show that we each have less than 1 chance in 50,000 of being killed by a global asteroid impact per 100 years. This is good news, because in 1995 the chances were 1 in 20,000.[26]

The downgrade does not mean that there are fewer giant Earth-crossing asteroids than we once thought, just fewer that are unknown. Since 1995, astronomers have discovered a larger fraction of the largest, most deadly asteroids and have determined that none are on a collision course with Earth. Calculations about the destructive effects of a tsunami resulting from an ocean impact have also been revised downward. But not all astronomers agree with Chapman's assessment, and until all planet-killing asteroids are accounted for, some still hold with odds that fall somewhere between the two extremes.

But why even consider the odds at all? Huge impacts are extremely rare, and even by Chapman's own rankings, the odds that you will be killed by a more frequently occurring event like the one at Tunguska are even less likely—1 in 5.7 million. In short, the explanation is this: an impact event by an asteroid of 3 kilometers (1.8 miles) or more is qualitatively off the scale compared to even the largest of the known world disasters, such as the Indian Ocean tsunami in December 2004. All of the greatest natural disasters, like tsunamis, earthquakes, typhoons, and volcanic eruptions, are geographically limited. Terrible though they may be, they are "contained" and hence not likely to trigger a global catastrophe. Moreover, there are a number of ways that an asteroid can, depending on the magnitude of its impact, kill people, either immediately or later. If the air-blast overpressure doesn't get you right off, the secondary effects will. These include, but are not limited to, tsunamis, firestorms, earthquakes, poisoning of the biosphere, climate change, drought, plagues, and famine.

Of all the cosmic interventions described here, an asteroid or comet impact of some magnitude is the one more likely to occur within a few centuries or so. For us, that means probably not tomorrow. Nor is the next impact, whenever it occurs, likely to be a civilization killer. The really big ones only come around every few hundred million years. But just looking at Earth's crater record, both recent and ancient, at the very least we could be subject to another Tunguska-style explosion, and perhaps next time in a more populated area. The consequences of impacts of various magnitudes are predictable, thanks to all the nuclear bomb tests conducted around the world. But the larger impacts are completely unique in their catastrophic potential—not just in terms of mortality and physical damage, but also in terms of the effects they could have on social order, economies, communications, and infrastructure, regionally or worldwide.

The threat is real enough for even some politicians to take seriously. In 2005, the US Congress passed a bill authorizing NASA to search for asteroids as small as 140 meters that could possibly strike the Earth. The directive also instructed scientists to analyze possible means of diverting objects on a likely collision course with Earth. It was nothing less than a mandate to defend the planet against errant asteroids and comets.[27] The bill, however, provided no money to accomplish either of these goals. In March 2007, planetary scientists met in Washington, DC, to highlight current capabilities in detecting NEOs and to discuss possible ways of deflecting or mitigating a threatening asteroid. Without forthcoming funding, however, there was little else they could do than trot out ideas. In his keynote address, Simon "Pete" Worden, director of NASA's Ames Research Center, said that the cost of finding at least 90 percent of the 20,000 estimated potential Earth-killers by 2020 would cost about $1 billion. Worden was clear: "We know what to do, we just don't have the money."

Assuming scientists had the money, what would they do? A white paper from the conference spells things out pretty clearly.[28] Here are a few recommendations:

- Funding should be increased for searches for asteroids between 140 meters and 300 meters. Objects this size are big enough to be hazardous, but not big enough to be spotted at great distances, and that means they can approach undetected and hit the Earth with little or no warning. If a 300-meter object struck in a populated area, it would release 2,000 megatons of energy and create a crater at least 5 kilometers across. If a similar-size object struck the ocean, it would generate a tsunami equivalent to the 2004 Indian Ocean tsunami. In either case, the impact could kill tens of thousands of people outright, injure many times that, and cause catastrophic property and infrastructure damage.

- Support the operation of Earth-based facilities such as the Arecibo radio telescope, which, because of funding issues, may be shut down in 2011. This giant 305-meter (1,000-foot) telescope, the largest single-dish telescope in the world, plays an essential role in refining the orbital tolerances of potentially hazardous objects, such as Apophis, and would be invaluable in providing basic information once a deflection technique is developed.

- Identify viable methods for deflecting Earth-bound asteroids and design a workable deflection campaign. This could be facilitated with an "Impact Response Exercise." Agencies responsible for assisting populations affected by natural disasters already conduct tabletop exercises to help them identify issues and develop solutions to help them respond more effectively to various emergency scenarios; a similar exercise should be designed for an asteroid impact, both on land and sea.

- Develop an international protocol for dealing with the threat and disaster "mitigation" of an asteroid or comet so as to minimize confusion, suspicion, and uncertainty among world governments.
- Develop a strategy for educating public officials on the hazards of Near-Earth Objects.

The white paper offered many more recommendations, all of them sound and all of which point out what a socially complex and technically challenging situation humanity must deal with if such an eventuality ever presented itself. Fortunately, planetary scientists are no longer merely talking about asteroid and comet threats and calculating what-if scenarios of impacts of NEOs; they've taken it upon themselves to toss out a few ideas as to how to deflect or otherwise mitigate a threatening asteroid.

Usually the first idea that leaps to mind is to detonate a nuclear device on or near the asteroid's surface. But this blow-up-and-pray approach may not be the best. The surface of an asteroid is rough and unconsolidated to some extent. Many asteroids have one or more companions. Others are so loose they are considered "rubble piles." If you blow up something like that, you would likely end up with numerous large asteroidal chunks following the same trajectory, like a cosmic shotgun blast.

How about docking a spacecraft with the asteroid and

pushing it out of the way? Sounds good, but there are several problems with this scheme, in addition to the two mentioned above. How do you dock with a rubble pile or an asteroid with multiple bodies? Moreover, most asteroids rotate, so an engine attached to the surface would be thrusting in a constantly changing direction. Stopping the asteroid's rotation, reorienting the spin axis to one that is more favorable, or firing the engine only when it's aiming in a certain direction would add undue complexity and waste time and propellant.

Instead of blowing it up, veteran NASA astronauts Edward T. Lu and Stanley G. Love suggest towing it out of the way with a robotic spacecraft they call a "gravitational tractor." Essentially, the added mass of the spacecraft, parked near the asteroid, would mutually attract the asteroid's mass and, over time, divert its path. To alter the course of a 200-meter asteroid, the tugboat would have to be on the order of 20 tons and positioned at about one-half the asteroid's radius above the surface. So for an asteroid 200 meters in diameter, the separation distance would be 50 meters. To balance the gravitational attraction between the two bodies, the spacecraft's engines would thrust away from the surface, but not so much that it would escape the asteroid's gravitational field. To prevent blasting the surface, the thrusters would be canted away from the asteroid body.[29]

There are two reservations about this approach. First, it can't be accomplished in a few weeks or months. A 20-ton gravitational tractor would need a lead time of about 20 years to effectively change the trajectory of a 200-meter asteroid. Larger asteroids would require more massive spacecraft, longer hover time, or greater lead time. Second, no mitigation technique, no matter how reliable, will be effective if we can't find an asteroid to mitigate, and, as stated, astronomers think there may be thousands of objects of this size out there that have yet to be detected. If one is discovered heading in our direction tomorrow, with a lead time on the order of months, obviously the gravitational tractor option would be out and the nuclear (desperation) option would be in.

Other innovative proposals call for nuclear detonations to nudge the asteroid off target, anchoring tethers onto the surface so that a spacecraft can gently sling it athwart, attaching sails to allow the solar wind to blow it off course, and gas-blasting the surface with a hybrid rocket motor at minimal thrusting—a kind of leaf-blower solution. Unfortunately, each of these options presents challenges similar to those described above.

For the really big asteroids for which there are long lead times, like centuries, some astronomers suggest changing the asteroid's surface thermal conductivity. The irregular and motley appearance of an asteroid's surface causes sunlight to reflect more strongly from some parts than others. As the thermal photons leave the surface, they carry momentum away as well, which produces a slight reaction force on the asteroid. This phenomenon, called the Yarkovsky effect, was first directly observed in 2003 by a team of astronomers using the Arecibo radio telescope.[30] But even before that, some scientists proposed using the effect to change an asteroid's direction by modifying the few upper centimeters of its surface, perhaps by blasting away loose surface material or coating the asteroid with some sort of material, such as soil. However, blanketing a 1-kilometer asteroid 1 centimeter deep would require a lot of dirt—about 250,000 tons of it.[31]

When it comes to deflecting NEOs, it seems there are few plans to deal with comets. Some of the proposed mitigation techniques might apply, but when you get down to it, comets don't lend themselves to any one mitigation technique. They are larger (0.5 to 12 kilometers, or 0.3 to 7 miles) than the 140-meter-range asteroids that astronomers talk about deflecting, they move faster than asteroids, and they have a crumbly composition that would be difficult if not impossible to fetter or flog. Comets are "friable"—that is, they are fragile and can easily break up, as was seen with Comet Shoemaker-Levy 9, which pummeled Jupiter in several fragments in July 1994. But comets have other issues as well. Those with known orbital periods in the inner solar system get "sandblasted" by solar winds each time they approach the Sun. The sandblasting creates a dusty trail of millimeter-size particles in the comet's wake. If the wake intersects with Earth's orbit, we see a harmless meteor shower when we pass through it. But sometimes the gravitational tug-of-war between the giant planets and a sizable comet can cause it to crumble, creating a train of rubble that can contain larger meter-size objects, and if that train crosses Earth's orbit, you have a potential hazard.

One such train, called the Taurid Complex, crosses Earth's orbit in the last few days of June each year. There is strong evidence that it contains many small comets and asteroids and boulder-size objects. The Tunguska event, which occurred on June 30, 1908, is thought to have been the result of a Taurid Complex comet fragment.[32] If Earth encountered a closely spaced chain of Taurid Complex fragments, the result would be many simultaneous Tunguska events. This scenario, in fact, is what some scientists think would be the most likely to occur.[33] How are we to mitigate against that?

As planetary scientists have been exhaustively reminding politicians and the public for decades now, whether society is prepared for it or not, a collision between Earth and an asteroid or comet is not an "if" but a "when." On the other hand, such a prediction doesn't cast a very big shadow when seen in the light of the apparent crater record on Earth. As cited earlier, there are some 174 known impact craters on the Earth, as well as a handful of unconfirmed sites, and the number is growing by the year. Their statistics are telling and illustrate the variety of extraterrestrial objects that have struck Earth in the past. The largest known impact structure is 300 kilometers (190 miles) in diameter and is located in Vredefort, South Africa. The smallest, in Haviland, Kansas, is only 15 meters (50 feet) across. The oldest known crater is Suavjärvi in the Republic of Karelia, Russia. It is estimated to be 2.4 billion years old. The youngest is Sikhote-Alin, created during that meteorite spray over Russia in 1947.

As of September 2007, however, another meteor crater was added to the world's roster by a multiton chondrite that struck near Carancas, Peru, south of Lake Titicaca near the Bolivian border. Technically, it ranks as the smallest in diameter (12 meters, or about 40 feet) and, of course, the youngest in age. Unfortunately, as of this writing, the local authorities have refused to extract the meteorite, which lies below the water table and is, hence, under a pool of water.[34] It is estimated that the impactor had a kinetic energy of 0.3 kilotons of TNT.[35]

The crater with by far the most notorious reputation is Mexico's 170-kilometer (100-mile) Chicxulub, left behind by the meteorite that killed off the dinosaurs. Recently, astronomers traced the origin of the Chicxulub impactor to the breakup of an asteroid 170 million years ago in the asteroid belt, (298) Baptistina. The original body had a diameter of about 170 kilometers when it was struck by another asteroid about 60 kilometers (40 miles) across. The breakup created what is now known as the Baptistina asteroid family. Scientists estimate that 20 percent of the Baptistina rubble was knocked out of the asteroid belt and that 2 percent of this went on to strike the Earth. One of the Baptistina fragments may have also created the young lunar crater Tycho.

Clearly, when asteroids collide in the asteroid belt, as asteroids will, they can have a devastating effect down the road for Earth.[36]

Virtually every landmass in the world is pitted with craters of all sizes and ages. They have fallen anywhere and everywhere and show no preference to longitude, latitude, or hemisphere. But if you think about it, 174 is not a very large number of craters for a planet to have, especially one that's been around for over 4 billion years. There should be more, hundreds more—but where are they? Given that the Earth is largely an ocean-covered world, obviously, that's where most impacts occur. However, only a handful of seafloor craters are known. Over the eons, many have no doubt been erased by the constant recycling of oceanic seafloor while others, perhaps due to the size and composition of the impactor, trajectory, and ocean depth, didn't penetrate to the bottom. Only the largest, most destructive impact events, like Chicxulub, would have left notable scars on the seafloor.

A mostly water world such as ours, however, still can't account for the lack of extensive evidence for impacts on land. Perhaps no one has stumbled across them yet, or erosion or other geologic processes have made them unrecognizable. What's more likely, though, is that a significant number of objects that struck Earth in the past never actually "struck" Earth per se but, like the Tunguska object, exploded in the atmosphere. Hence, although no crater would be produced, the object's aerial burst could reach the ground with varying effect. Here again, a continuum of impactors and impact scenarios present themselves. One may flatten a forest and incinerate hundreds of reindeer, another may spark widespread grassland fires or turn desert sand to glass, and still another could conceivably precipitate a minor extinction event. In all cases, even without an obvious crater, evidence for these impacts could be found in the by-products created out of the resulting explosion and ejecta, such as glasslike spherules created in the blast, high deposits of iridium brought to Earth by the impactor, and terrestrial fallout (soot from organics).

Recently, intriguing evidence for the latter type of event was found in North America—one that, if true, has disturbing implications as to how often Earth truly suffers a cosmic occurrence of this kind. It bears on a short period some 12,900 years ago known as the Younger Dryas. At the heart of the matter is why the period of interglacial warmth following the last ice age was suddenly broken by colder temperatures, expanding ice sheets, and the marked advancement of mountain glaciers. The Younger Dryas lasted for only 1,000 years, but during that brief period megamammals like wooly mammoths, mastodons, ground sloths, sabertooth tigers, and the western camel disappeared from North America. Ice cores of the Younger Dryas layers show evidence of widespread wildfires across the continent that raged for at least 50 years. Moreover, the Paleo-Indian culture called the Clovis people, who had hunted and ranged across North America for over two centuries, themselves vanished. What happened?

James P. Kennett, a paleo-oceanographer at the University of California, Santa Barbara, is a member of a team of scientists that claims to know the answer: an extraterrestrial impact, and a pretty big one, caused the cooling and knocked off both the megamammals and the Clovis people. In the dry technical language of scientists, Kennett calls the impact "a major perturbation."

His team argues that evidence for the Younger Dryas impact comes in the form of a carbon-rich deposit known by the sinister-sounding name "black mat." It is present at more than fifty archaeological sites across North America, including nine known Clovis hunting sites. The black mat, which was rapidly laid down on top of the bones of extinct mammals as well as Clovis tools and fluted spear points, has been instrumental in providing paleontologists with absolute dates of various Clovis sites. At the base of the black mat, Kennett and his colleagues found discrete layers containing tiny magnetic balls rich in titanium, the rare metal iridium in concentrations well above background levels, charcoal, soot, stringy glasslike carbon associated with the melting of resins in conifers, and extraterrestrial values of helium and carbon. "There is no obvious way of interpreting this assemblage of evidence," says Kennett, "other than [as] an ET impact."

If the claim holds up, he notes, "it would be the first record of an extraterrestrial impact with widespread consequences for anatomically modern humans." At the time, the Clovis lived throughout North America and hunted many of the larger mammals. After the impact, however, the extensive wildfires would have burned away the vegetation that sustained the larger mammals, causing them to starve. The Clovis, finding themselves without a food source, would themselves soon have to choose between starvation or scattering southward to more habitable locations.

The impactor, the scientists think, may have been the low-density fragments of a comet that exploded above the surface, creating a multiple Tunguska-like strafing run. The brunt of the impact is thought to have occurred over the massive Laurentide ice sheet, once located in the northeastern region of North America. Heat from the explosions would have melted the ice and created abundant freshwater outflows into the saline Arctic and northern Atlantic surface waters. This, in turn, would have slowed or shut down the thermohaline current, which transports heat north from the midlatitudes, and induced a millennial-long period of cooler temperatures.[37]

Intriguingly, the scientists also detected black mat layers extending through the rims of 15 elliptical lakes and wetlands known as the Carolina Bays, hundreds of which are concentrated along the south Atlantic seaboard. As seen from a high altitude, or better yet, space, the Carolina Bays are uncannily oriented with their major axes pointing northwest, toward the Great Lakes. Could these be multiple impact craters from a comet or asteroid breakup that occurred 12,900 years ago? This theory remains controversial. For one thing, no shocked rock or shatter cones have been found that can be associated with the bays, so there is no evidence of them being created by an impact. For another, geologists' analysis of the stratigraphy and rim deposits of a sample of the bays shows that some of them date back as far as 74,000 years, significantly predating the Younger Dryas impact. Clearly, more research needs to be done on all fronts to settle the matter.

Given what we know about the kind of kinetic energy an asteroid or comet impact can generate, it doesn't take a great mental leap to imagine the effects such an event would have on society and the infrastructure in general. As we've seen, the blast of a nuclear weapon can be on par with that of a small asteroid impact. Hence, we might base our scenarios on the apocalyptic consequences a nuclear war would have on society. In the late 1970s, the US Senate Committee on Foreign Relations requested the Office of Technology Assessment to produce a report that would make the general public better understand "the effects of nuclear war on the populations and economies of the United States and the Soviet Union." The OTA's assessment, published in May 1979, was titled simply *The Effects of Nuclear War*.[38] It is still a hair-raising read, particularly for someone like myself who was raised in the "duck-and-cover" days. In an effort to make nuclear war and its aftermath less of an abstraction, however, the authors included an appendix called "Charlottesville: A Fictional Account by Nan Randall." In Randall's account, 100 million

American citizens are killed outright in the nuclear exchange, but Charlottesville, Virginia, not being a category 1 military target, is miraculously spared a direct hit. The narrative thus provides readers with a vicarious viewpoint of what problems and situations survivors might have to face after the bomb. Needless to say, Randall cannot avoid describing some rather unpleasant scenes, even though she spares readers visions of anarchy, social disintegration, and the imposition of martial law.

The Effects of Nuclear War has long been consulted by scientists who have wondered about how a nuclear explosion would compare with an asteroid or comet impact. The effects of each are, it turns out, chillingly similar, with both events producing an intense flash, an electromagnetic pulse that would induce damaging current and voltage surges, a fireball hot enough to generate instantaneous fires, and a highly destructive, death-dealing blast wave. One scientist who has probably made the most use of *The Effects of Nuclear War* is James A. Marusek, a nuclear physicist and engineer with Sandia National Laboratories. Marusek presented a paper at the 2007 Planetary Defense Conference called "Comet and Asteroid Threat Impact Analysis" that draws heavily on this unique government report.[39] In his treatise, Marusek describes in great detail a worst-case scenario: the effects of an asteroid 5.8 kilometers (3.6 miles) in diameter crashing nearly head-on into the Atlantic Ocean 750 kilometers (470 miles) northeast of Miami in the middle of June at 9:45 p.m. eastern standard time. He divides the outcome of the impact into five main effects: shock wave (atmospheric blast wave, ground shock, and tsunami); thermal radiation (flash and fireball); debris and aerosols (ejecta debris blasted into the upper atmosphere and space); electromagnetic effects (electromagnetic pulse, electrophonic bursters, and ionizing radiation); and secondary effects (mass fires, earthquakes, landslides, volcanoes, dust, and nuclear winter). He then assesses these effects from various locations, including Bermuda; Washington, DC; New York City; a town outside Indianapolis; Chicago; Dallas; Lincoln, Nebraska; and Los Angeles. In addition, and with a nod to Nan Randall's fictional account, Marusek provides his own vivid fictional renderings for Bermuda, Washington, DC, Indiana, and Los Angeles.

Obviously, locations nearest the point of impact would be most severely affected. Hence Bermuda and Washington, DC, which are about 1,000 and 1,200 kilometers (620 and 745 miles) from the impact site (72° 49' west, 28° 0' north), respectively, will experience the most devastation. Here is what Marusek predicts will happen at these two locales.

The kinetic energy released by the asteroid impact would be on the order of 10 million megatons. (The largest nuclear bomb ever tested, Russia's "Tsar Bomba," had a yield of 50 megatons.)[40] At Bermuda's distance, witnesses who would happen to face the direction of the impact would see a brief but brilliant flash over the southwestern horizon, and some would hear loud snapping or crackling sounds (the so-called electrophonic bursts thought to be caused by very low-frequency radio waves produced upon impact). The island's power grid and all other electronic and electrical devices, including automobiles and communications, would instantly go dead. Thirty seconds later, a fireball would sweep across the island for five minutes, causing most combustibles to ignite and burn. In the middle of the fireball's onslaught, a succession of violent ground shocks would level most buildings. Five minutes after impact, falling dust and debris from the sky would engulf Bermuda, making it difficult to breathe. Those still alive would see the fireball brilliantly backlit on the sky by the dust, creating an otherworldly scene. Another devastating aftershock, this one reflected off the Earth's solid core, would arrive twenty minutes after impact. Twenty-two minutes later, the thunderclap of the impact would finally resound across the sky. Four minutes later, an atmospheric blast wave with an overpressure of at least 10 pounds per square inch (psi) and a peak wind velocity of 467 kilometers per hour (290 miles per hour) would instantly flatten most buildings and kill most of the people. The coup de grâce would occur an hour and a half after impact with the arrival of a tsunami half a kilometer (0.3 mile) high.

What about Washington, DC? You would have to be standing atop the Washington Monument to see the silent flash of the impact. Immediately, the electromagnetic pulse would knock out the power grid, automobiles, emergency generators, and most electronics and computers. The ionizing radiation would jam communications for 10 hours. Thirty seconds after the initial flash, the fireball would lay into the city, igniting combustibles and triggering numerous fires. Even the pavement would soften enough for tire tracks to leave depressions. Violent ground shocks would soon follow, and the airborne debris would provide a backdrop to the fireball, revealing it in all its apocalyptic glory. Not until an hour after impact would citizens hear the sound of the impact as a series of echoing thunderclaps. The blast wave would hit the town 42 minutes later with an overpressure of 5 psi and a top wind speed of 257 kilometers per hour (160 miles per hour). This would be enough to blow houses off their foundations, severely damage sturdy buildings, and kill hapless victims with flying debris. If the president were still at the White House, he or she would best get out of the city immediately, because just two hours after impact, a tsunami 0.46 of a kilometer (0.3 of a mile) in height would sweep across the capital city resulting in its total destruction.

New York City? Destroyed by tsunami. No survivors.

Small town outside Indianapolis? The ground shock would trigger large earthquakes on the New Madrid fault system, causing significant damage to structures, dams, roads, and infrastructure. Gas lines would rupture. Fires would spread due to the inability of the fire department to respond. Following these events would be starvation, contaminated drinking water and food supplies, and disease, although not necessarily in that order.

Chicago? The ground shock would produce a small tsunami in Lake Michigan that would swamp the shoreline. Acid rain, black rain, and large violent storms would menace the region for months, while the population would be weakened by starvation and contaminated food and water supplies.

Dallas? Big D would lose its power grid and experience protracted violent storms, acid rain, starvation, sickness, and disease. The New Madrid earthquake would also cause minor structural damage.

Lincoln, Nebraska? Ground shocks would awaken the Yellowstone volcano (the largest in North America), creating a catastrophic eruption that would spew forth lava, ash, and a variety of gases including sulfur dioxide into the atmosphere. The ash and aerosols would contribute to nuclear winter conditions in the Northern Hemisphere.

Los Angeles? A large earthquake on the San Andreas Fault system would cause major structural and infrastructural damage throughout southern and central California. The quake would rupture a major dam in the San Fernando Valley. And a huge landslide on the Big Island of Hawaii caused by ground shocks would send a tsunami barreling toward the Pacific Coast, drowning the shoreline and driving 50 kilometers (30 miles) inland.

A rather bleak assessment, to be sure. It vividly illustrates how much more global such a disaster would be than, say, a major earthquake or a category 5 hurricane. After such an event there would be no one watching CNN and thinking "I'm glad it's not me," because there would be no CNN to watch. (No doubt the network would bounce back soon enough.) But as I said, this is a worst-case scenario, and as I also mentioned earlier, astronomers aren't likely to overlook a large killer asteroid or comet moving about in the Earth's neighborhood. If one were heading our way, we could assess the probability of an encounter over the next century, probably, and deal with the threat.

It's hard to get your arms around an eventuality that is not only highly unlikely but inconceivable. Okay, you might say to yourself, a catastrophic impact may have happened in the remote past and it may happen again in 100 million or 200 million years. But who can imagine such a thing happening now or tomorrow, or say on a cool October night beneath a full moon while watching your kid play in a high school football game? The probability of such an occurrence lies more in the realm of imagination than in fact.

But we might just as well ask who could have imagined a comet breaking up into a "string of pearls" and plunging into Jupiter's upper atmosphere in July 1994. How improbable was that? There were scientists at the time who predicted that nothing would happen, that the fragments of Shoemaker-Levy 9 would have no more effect than a fly striking the windshield of a bus. And then, when we saw the dark, ring-shaped blisters raised in Jupiter's cloud bands by the impacts, there was much marvel and wonder and heartfelt talk about how rare such a spectacle was and how lucky we were to be alive at this time in the history of civilization to witness it. Indeed, how lucky we were.

CHAPTER 5

KEEP YOUR DISTANCE

I conclude, therefore, that this star is not some kind of comet or a fiery meteor . . . but that it is a star shining in the firmament itself—one that has never previously been seen before our time, in any age since the beginning of the world.
    —Tycho Brahe, *On a New Star*

In his short story "The Star," written in 1954, Sir Arthur C. Clarke tells the tale of a crew of space travelers returning to Earth following their exploratory mission to a supernova remnant. Still orbiting the stellar cinder is a single planet that had been seared by the conflagration 6,000 years before. On the surface they make a disturbing discovery—a vault containing the record of a race of advanced beings that was still flourishing when their sun exploded, destroying all life. From the astronomical evidence, the astronomers are able to precisely determine the date of the explosion and in turn determine when the light from the supernova would have reached Earth. The final paragraph delivers the ironic punch line:

There can be no reasonable doubt: the ancient mystery is solved at last. Yet, oh God, there were so many stars you could have used. What was the need to give these people to the fire, that the symbol of their passing might shine above Bethlehem?

Poor God gets the blame for so many of our calamities, both natural and self-made, but the fact is stars *do* blow up quite often in the universe without the aid of divine intervention. You might say it's in their programming. Stars that are eight to ten times more massive than the Sun live far shorter lives than stars like our own—on the order of a few hundred million years for the lighter-weight stars but only a few million years for the most massive species—and at the end, they explode violently.

In general, they can go out in one of two ways, as a thermonuclear explosion or as a core-collapse supernova. Briefly, the thermonuclear variety is thought to occur when an extremely dense white dwarf star accretes too much matter from an otherwise normal companion. At some point the accumulating matter becomes massive enough to abruptly "restart" the fusion process, and the whole star goes up in a single thermonuclear *whoosh!* leaving nothing behind except elemental ashes. The second type, a core-collapse supernova, involves a single massive star that has fused all of its elements—hydrogen, helium, oxygen, carbon, and so on—until only iron remains. This is where the star hits the proverbial wall because the amount of energy needed to fuse iron is greater than the fusion process can provide. Hence, the nuclear fusion ceases and the iron core implodes. The implosion subsequently rebounds in a titanic explosion, blowing the outer layers to smithereens. What's left of the core may, depending on the mass of the progenitor, be either an extremely dense, Earth-size object composed mostly of compressed neutrons—a neutron star—or, smaller and denser still, a black hole.

Whether the star's demise is via the thermonuclear route or core-collapse, the resulting blast expels most, if not all, of the star into space. The expanding cloud of debris, called a supernova remnant, may extend up to 5 or 10 light-years within a few centuries after the explosion, sweeping up interstellar gas and dust, heating the interstellar medium, and swamping any stars in its path. Eventually, the remnant's shock waves slow, and the gas cools and disperses into the interstellar medium. The whole process takes about 100,000 years, although there is evidence for the gas shells of remnants as old as several million years and covering a radius of about 6,000 light-years.[1] Astronomers base the frequency of supernovae in the Milky Way on those observed in other similar galaxies, as well as the dates of known supernova remnants in our part of the Galaxy. The numbers vary a little depending on who you talk to, but in general astronomers predict two to three stars blow up every century in galaxies just like ours.[2]

Now imagine for a moment that the number of civilizations in our Galaxy is of *Star Trek*ian proportions—that is, in the millions. If there are, say, two supernovae per century on average, then there could have been at least 20 million supernova explosions somewhere in our Galaxy over the last billion years. One might assume, then, that more than a few civilizations or potential civilizations may have been wiped out or adversely affected over the last billion years by exploding stars. But 20 million supernovae amount to only 0.01 percent of the 200 billion stars populating the Milky Way—thus ensuring that a civilization-killing supernova must be a fairly rare event (though keep reading for an exception). And certainly your typical massive supernova-producing star, with a mass ten times that of the Sun, could never host a planet with intelligent life because it lives only a few tens of millions of years. It took the Earth-Sun system at least a billion years to form, and another billion for the first simple bacteria to evolve. That's already about 200 times the life span of a massive star, never mind that sentient life takes another couple of billion years beyond that to develop. Hence, it's a safe bet that intelligent life, even sub-intelligent life, is not likely to be found on planets orbiting such super suns.

But—here's the exception—earthlike planets can hatch in quiet backwaters of the Milky Way only to, at some point in their history, find themselves in more perilous parts of the galactic neighborhood. If all the stars were pinned in place in the Milky Way, this wouldn't be an issue, but the stars aren't pinned in place, they move about in orbits around the center of the Galaxy. Not only that, they bob up and down in the galactic plane like a skiff in a choppy sea. From Earth, if we could reduce several million years of these motions to a minute's time, we would see that all stars move with greater and lesser speeds. Like Van Gogh's *Starry Night*, the heavens would be streaked with arcs of stars racing across the sky as if they were meteors. Thus, in the real estate that is the Milky Way, stars that were once far away can trespass into another star's buffer zone over periods of several millions of years. The timing would have to be right. But if a star that is likely to go supernova ends up near a star harboring a planet—and on which a civilization is beginning to emerge or is in flower—then there *would* be a danger of atmospheric damage, drastic climate change, genetic mutation, or even total extinction by a supernova. A cluster of such massive stars treading within the buffer zone could double or treble the danger by increasing the number of potential supernovae. It's not outside the realm of possibility. And as we shall see, this may have happened to Earth millions of years ago.

It goes without saying, then, that any civilization that aspires to some form of sentience should keep its distance from such volatile stars for periods of at least several billion years, longer if said civilization aspires to travel between the stars. How close is too close? Astronomers aren't absolutely certain, but they have a pretty good idea. A supernova's danger zone depends on a number of factors, including the star's original mass, the direction its radiation travels in space after the explosion, and the duration of its radiation. Back in 1995, astronomers concluded that a nominal buffer zone is a radius of about 32 light-years, but that's not a hard-and-fast rule.[3]

Would it be tempting fate if a supernova explodes at, say, 50 or even 40 light-years away? It's fascinating to consider the myriad scientific options addressing this subject because a lot of what might happen depends on the planet, too—particularly its atmosphere. Fortunately for us, though, this is largely an academic question. Technically, the nearest supernova candidate is IK Pegasi, a binary star system in the constellation Pegasus. The primary is a hot star nearly five times the diameter of the Sun but only about twice its mass.[4] I mention it because some scientists seem to think it could go out as a supernova, while other experts are less certain. At a distance of 150 light-years, it would certainly be bright, but probably not dangerous.[5]

The nearest true supernova candidate is also a binary star system, Alpha Crucis, which forms the southern tip of the Southern Cross. It lies 320 light-years away and hence is well outside our buffer zone. The second-nearest true supernova candidate, Betelgeuse in the constellation Orion, lies somewhere between 500 and 600 light-years away. It's already in an advanced stage of evolution and is expected to go supernova in about 1.5 million years.[6] Both Alpha Crucis and Betelgeuse, when they explode, will be nearly a billion times brighter than the Sun and from their distances will appear as bright as the full Moon in the sky. Future generations need not worry about these stars wandering into our neighborhood because the Sun will not be in the vicinity of either by the time they go supernova.

The branch of astronomy charged with keeping tabs on all this chaotic stellar motion is called astrometry, or positional astronomy. Eyeball astrometry got rolling in 1718 when Edmund Halley noticed that three bright stars—Sirius, Procyon, and Arcturus—were about half a degree away from their position as plotted more than 1,800 years earlier by the Greek astronomer Hipparchus. Precision astrometry, however, is not a very old science, because measuring how much a star changes in position with respect to the more distant background stars is extremely difficult. Even the nearest stars, which show the largest motions on the sky, move only a few arcseconds a year at most. That's a very small amount. One arcsecond is about the size of a quarter seen 5 kilometers (3 miles) away. The motions of more remote suns are measured in *milli-arcseconds* per year. That task is equivalent to measuring the width of a quarter thousands of times more distant. A giant leap forward in the field was based on the advent of the astrometric satellite called Hipparcos, which, in the late 1990s, pinned down the positions of 118,000 stars to within an accuracy of 0.002 arcsecond, a feat impossible from the ground.[7]

It was astrometry, in fact, that gave astronomers their first hint that our peaceful part of the Milky Way had once been ravaged by supernovae. Astronomers measuring the motions and distances of stars in the solar neighborhood noted that the Sun appears to lie along the edge of a "pocket" in the Galaxy, called the Local Bubble, that is devoid of bright stars. Remarkably, this stellar hollow is situated at the intersection of the galactic plane and a vein of stars inclined about 20° above and below the plane. This inclined plane, known as Gould's Belt, is thought to be an appendage extending from a spiral arm of the Galaxy, though other astronomers argue that it may not be a coherent structure at all but rather a chance alignment of several smaller star groups with no common origin.[8] Whatever the belt's true nature, astute star gazers can see it for themselves by noting the number of bright stars lying in the direction of the constellation Orion in the winter sky and those in and around the constellation Scorpius in the summer. This latter region of Gould's Belt contains a loose assemblage of hot, massive stars called the Scorpius-Centaurus association. It is a breeding ground for supernovae.

Beginning around 1995, astronomers began earnestly putting their models of how a nearby supernova might affect life on Earth together with observations of the solar neighborhood. The idea, however, goes further back than that. "The suggestion that super-novae or something similar has caused extinctions dates back to at least the 1950s when German paleontologist Otto Schindewolf suggested it as an explanation for the end-Permian mass extinction," says Doug Erwin, senior scientist and curator at the Smithsonian's National Museum of Natural History in Washington, DC. "Since then it has been regularly resurrected, usually without the authors knowing how often it has been previously proposed." The only way of putting such suggestions on a scientific basis, he says, "is to find a distinctive marker, as with iridium and the impact of an extra-terrestrial object for the KT [Cretaceous-Tertiary boundary]."[9]

That is exactly what astronomers did in 1999 using the technique called accelerator mass spectrometry. When your typical supernova explodes, it produces an isotope of iron called ^{60}Fe in quantities ten times the mass of the Earth; only negligible amounts of this form of iron are produced within the solar system. In an analysis of 13 million years' worth of ocean-crust layers, astronomers found enhanced concentrations of ^{60}Fe—as much as two times the normal amount—in three layers dating back some 3 million years. The iron wasn't just sprinkled on, either; it was dumped. Just as iridium can only be deposited in great quantities by asteroids, that much ^{60}Fe had to have come from the stars. This, then, was the smoking gun of a supernova explosion between 2 and 3 million years ago and less than 100 light-years away.[10]

It has also been proposed that the Local Bubble was carved out by not just one supernova but several within the last 10 million years—supernovae most likely affiliated with the notorious Scorpius-Centaurus gang. The Bubble, which is about 500 light-years across, is surrounded by at least two super shells of gas from the remnant, or remnants, that are still aglow with X-rays produced by the dead stars. The innermost shell, the Local Bubble itself, contains hot, rarefied gas and only a few stars. It's as if the region has been sterilized of potentially lethal galactic hazards. Along one edge of this quiet interstellar sanctum the Sun abides like a pearl in an oyster.

But there's more. Astronomers think the Scorpius-Centaurus association could have yielded as many as twenty supernovae in the past 10 million years, all within the Sun's vicinity. And 2 to 3 million years ago, the center of the Scorpius-Centaurus association would have been at its closest, about 320 light-years away, but the outlying members of the association could have been between 90 and 130 light-years away. That's still outside the buffer zone but close enough to have deposited all that radioactive iron and create some serious secondary effects. An intense influx of supernova-generated cosmic rays lasting for about 100,000 years could have eaten away at Earth's ozone layer. This would have allowed greater amounts of ultraviolet radiation from the Sun to reach the surface. And this could have possibly contributed to the biodiversity crisis known as the minor Pliocene-Pleistocene marine extinction.[11] It coincided with the time when the Scorpius-Centaurus association was making its closest approach to the Sun. More ultraviolet radiation in the tropical seas (due to the high solar angle) would have killed off plankton, which in turn would have meant fewer numbers of mollusks and other marine invertebrates—one of the traits of the Pliocene-Pleistocene extinction. Though only a theory, it's considered plausible enough by some geologists and paleontologists.

But that was 2 million years ago. What about tomorrow? Could the Sun one day stray too near a potential supernova, or vice versa? Seth Redfield, an astronomer at the University of Texas at Austin who studies and models the local regions of the interstellar medium, says that's problematic. "This has been something people have been looking into," he says. "There are certainly candidates with rough ages, but none are particularly solid. The Hipparcos mission has provided information on the recent past encounters that the Sun has likely had with nearby stars, but it only provided accurate distances to the very nearest stars, so it could only probe back over 10 million years and couldn't take into account stellar evolution." It will take a larger astrometry mission, like the planned European mission Gaia, to better determine the Sun's interaction with stars in the past and the future. Gaia will measure the positions of one billion stars during its five-year mission, slated to begin sometime in 2012. We will have to wait for our answer until then.[12]

What about a potential supernova that may be lurking nearby, hiding behind dense clouds of interstellar dust? This is unlikely. The many multiwavelength surveys of the sky, particularly at infrared wavelengths that can "see" into otherwise impenetrable shrouds of dust, would probably have revealed such a menace long ago. We can conclude, then, that, although supernovae may have caused some havoc millions of years ago, just as the first hominids began moving upon the African savannahs, they won't again be a player for another million years at least. If you are as optimistic as I am about the fate of our civilization, then you are probably not overly worried about this.

But though we may be safe in time, we may not yet be safe in distance. Recall that the power of a supernova depends, among other things, upon the progenitor's mass. The more mass a star has when it blows, the greater the energy released. But even supernovae have their limits. Astronomers are now aware of cosmic objects that erupt with many times the energy of the most powerful supernova. How much bigger would Earth's buffer zone have to be to avoid such an object? Like most astronomical realities, the answer is: it depends.

Visible in the southern constellation of Carina the Keel is an inconspicuous star called Eta Carinae. To see it, you need binoculars or a small telescope—and even then it's not very bright. But Eta Carinae can hardly be called an unassuming star. It is, in fact, extremely unstable and has a mass greater than 120 Suns. At a distance of more than 7,600 light-years, it is the nearest supermassive star known. It is also a rarity. Our Galaxy probably has only a dozen or so stars that are as massive or more massive. When *this* monster goes—and it could go at any time—it will be a beauty, much brighter than Venus and easily seen during the day. And Earth will have a front-row seat to a stunning supernova drama, though, thankfully, not so close as to put us in danger.

At least that's the assumption. But to date, Eta Carinae has defied astronomers' assumptions about its nature and behavior. A few centuries ago it was a star of middling brightness, like those

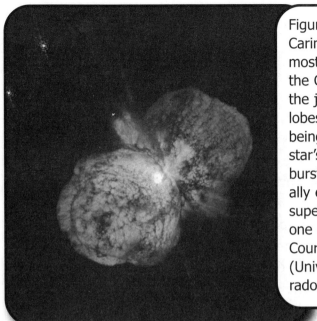

Figure 11. Eta Carinae, one of the most massive stars in the Galaxy, lies near the juncture of two lobes of gas and dust being ejected by the star's energetic outbursts. It will eventually explode as a supernova, but no one can say when. Courtesy of Jon Morse (University of Colorado) and NASA.

in the Little Dipper. Then, beginning in 1837, its brightness increased until for a few years it rivaled Sirius, the brightest star visible in the night sky; its color became tinged orange-red.[13] Today, Eta Carinae has once again dimmed to the point that it's not visible to the naked eye unless you use binoculars or a telescope. That doesn't mean it's quieting down, though. In fact, this lull may be the proverbial calm before the storm.

In 1998, the star suddenly doubled in brightness, becoming nearly visible to the naked eye. In addition, the nebular shells surrounding the star increased in brightness by 30 percent.[14] Astronomers were caught flat-footed. "Its brightness increase surprised us for a couple of reasons," said Kris Davidson of the University of Minnesota at the time. "First, we thought we had good reasons not to expect a major outburst in the next few decades. Second, when an eruption does occur, [the star's] spectrum should change, and its color should get cooler. Both are contrary to the data."

Davidson, who has studied the troubled star for years, still marvels at Eta Carinae and its reputation for being "the most extreme object known in our part of the Galaxy. We think it began with a mass above 150 solar masses, maybe 180. It now radiates about 5 million times the Sun's luminosity. No other star, anywhere, has been proven to be more extreme. It will probably become a supernova, maybe a hypernova, in another 100,000 years."[15]

Hypernova! Well, astronomers can't very well call such a phenomenon a "super-supernova," but in principle that's what a hypernova is. This relatively recent addition to the astronomical cabinet of wonder is one of the most powerful of all known stellar explosions. The smaller hypernovae would be about 100 times more powerful than the most powerful supernovae. Some astronomers speculate that Eta Carinae is the nearest best candidate for achieving hypernova-hood. If true, then we must take another look at Earth's buffer zone.

The first thing we must ask ourselves is how massive Eta Carinae might be when it finally explodes. A star like Eta Carinae is a huge star, and it takes a long time to shed tens of solar masses, longer than its lifetime of a few million years. What keeps it from shedding mass in great quantities is the struggle all stars have between gravity, which wants to collapse the star, and radiation, which wants to expand it. The star wants to collapse and it wants to expand but it can do neither, and so it remains in this unremitting, one might even say existential, balance that astronomers call hydrostatic equilibrium.

But an astrophysical process going on in the tenuous upper layers of this mighty star's atmosphere acts like a valve, gradually letting the air out of the balloon, so to speak. The thermal energy in this region, for reasons we need not concern ourselves with here, is actually millions of times hotter than the surface of the star itself. This heat creates a "super wind" that overwhelms gravity and blows ionized clouds of electrons and protons into space. This is how a big, bad star loses mass over time. Usually, the process is continuous and regular, and the mass loss steady. In Eta Carinae's case the star loses somewhere between 0.001 and 0.0001 of a solar mass each year.[16] But then there are the rare times when the star convulses and coughs up gobbets of mass equal to two or three Suns into space, as it did in 1837. Over time, both the steady and the convulsive shedding of mass takes its toll. A star that begins its life as Mister Universe with a mass sixty times that of the Sun may, depending on its mass-loss rate, wind up as a beach weenie only four or five times the Sun's mass at the end of its life.

If the lower mass range of hypernovae falls between 20 and 30 solar masses, how much material must Eta Carinae shed to avoid blowing up as a hypernova? A little intuitive arithmetic tells you that if the star sheds mass at a steady rate of 0.001 solar mass a year, it will shed 100 Suns' worth in 100,000 years. By then it should be near the 20-solar-mass hypernova threshold (*if* its current mass is 120 Suns). But if the mass loss is closer to 0.0001, it will have only dropped by 10 Suns in 100,000 years and 100 Suns in a million years. Of course neither of these takes into account the mass loss due to eruptions, but one could speculate that they would, if they occur regularly, spit out another 20 solar masses in 100,000 years.

If, if, if. Is there anything certain about Eta Carinae? The star itself has never been observed directly because it is surrounded by the beguiling nebula astronomers refer to as the Homunculus.*

Homunculus, Latin for "little man," has all sorts of curious interpretations, some obscure, others grotesque. The first person to link this appellation to Eta Carinae was Enrique Gaviola of Córdoba Observatory, Argentina, in a paper published in the *Astrophysical Journal* in 1949. He describes Eta Carinae's nebulosity as having "a shape resembling a 'homunculus,' with its head pointing northwest, legs opposite and arms folded over a fat body."

No other object in the Galaxy better illustrates stellar bedlam in the making. It consists of two enormous lobes of gas and dust that lie on opposite sides of the enshrouded star, forming an hourglass that is tipped away from us. The star itself lies somewhere within the "neck" of the hourglass, brilliantly illuminating it from within. Surrounding the neck is a "skirt" of smeary-looking material that gusts out from the star's equator (or is, at least, perpendicular to the neck of the hourglass). Framing the entire hourglass is a nuclear fireball frozen in time, with cirrusy streamers of gas threading out into space. This outer effluvia, the remnants of past eruptions, extends out to 200 to 300 light-years, a volume that, if it is similar to our solar neighborhood, could contain several thousands of stars.

Although there are still many questions concerning Eta Carinae's distance and mass, and why the Homunculus nebula is shaped the way it is, astronomers have a good idea of the direction that the explosion would take should it become a supernova or hypernova. High-resolution spectral analysis of the motion of the gas within the Homunculus indicates that about 75 percent of the total mass and 90 percent of the kinetic energy were released at latitudes between 45° and the pole.[17] That and the star's axial orientation—41° away from the Sun—indicate that most of the expulsed radiation will be directed away from Earth. "Fortunately for us, we're not located along Eta's polar axis," says Davidson, "but thousands of planets in our galaxy probably are."[18] The reason: the star's polar axis is inclined a mere 10° from the star-crowded galactic plane and roughly in the direction of a spiral arm.

Obviously, any advanced civilizations percolating within the purview of Eta Carinae might consider relocating if they can. But how far should they retreat? That again would depend on whether Eta Carinae goes out as a supernova or a hypernova.

Given its probable mass at its climactic end, the former scenario would be dicey for any planet within 150 light-years. But if Eta erupts in a double-barreled hypernova fashion, something unique and diabolical will emerge from the conflagration, something astronomers are only now beginning to come to grips with.

It turns out that most of the energy of a hypernova is released in the form of gamma rays, which are X-rays pumped up to much higher intensities. The extra gamma-ray kick arises from the kinetic energy stored in a shell of particles (photons, electrons, and such) moving at near-light speeds in the fireball produced by the initial explosion. Like a cork being suddenly pulled from a bottle of champagne, as the fireball expands, the photon density thins enough to allow gamma rays to escape in a flash of radiation known as a gamma-ray burst. Subsequent shock waves caused by fast clumps of gas slamming into slower clumps produce a gamma-ray "afterglow."[19]

The point is that life-forms arrayed within this gamma-ray beam at various distances will be at risk of total destruction if they are too close—and genetic damage, mutations, and climatic effects will occur at progressively farther distances. The fact that bursts from the distant reaches of the universe can be detected by Earth-orbiting satellites testifies to the intensity of their emission.

Unfortunately, no one knows how far you'd have to be from a hypernova-turned-gamma-ray-burst to be safe. One study suggests that even from a distance of 10,000 light-years, a beamed gamma-ray burst could zap the surface of a planet with a thick atmosphere, like our own, with an enhanced dosage of ultraviolet radiation.[20] After all, if an event can be detected across billions of light-years of space, then a gamma-ray burst anywhere in the Milky Way could have unfortunate consequences, if not for us then for other potential life-forms. "In one scenario," says

Davidson, "almost all the communications, reconnaissance, and GPS satellites would go dead, all together, except for those in Earth's shadow. Imagine how confused the response would be!" A plot line worthy of Hollywood, to be sure.

It is the long-distance reach of a gamma-ray burst that makes it more potentially threatening than a supernova to life-forms. Assuming there may be a dozen or more stars in the Galaxy that are more massive than Eta Carinae, should we be worried? Most astronomers would say "probably not," again because of Earth's location in both time and space. Gamma-ray bursts come in at least two varieties—short-duration and long-duration. The former are rare in the solar vicinity and the latter, as we shall see, apparently don't thrive in an interstellar environment like the one in our Galaxy, and may even be on their way to extinction.

Short-duration gamma-ray bursts last from thousandths of a second to about two seconds, while the long-duration type may last up to twenty seconds or more. Both produce gamma-ray and X-ray radiation, but the short-duration variety produces "harder," more energetic emissions. The long-duration burst, however, lasts 100 to 1,000 times longer, and so even though its radiation is "softer," its staying power makes it potentially the more dangerous of the two.

It's a complicated business; the two types of bursts result from three, maybe four (maybe more!) different astrophysical processes, the details of which astronomers are still trying to work out. We've already discussed one, the long-duration burst resulting from the collapse of a supermassive star—at least 20 times more massive than the Sun—into a black hole. Such an object is a one-in-a-hundred-billion type of star in the Milky Way. Eta Carinae easily qualifies. But, as mentioned, if the star sheds too much of its mass before it collapses on itself and explodes, not enough residual mass will be left to produce the all-important black hole, whose jets can blast their way through the star's outer envelope and beam the concentrated energy across interstellar and intergalactic space. On the other hand, if it retains too much mass, the black hole may not be able to burn through the dense outer layers of the star. "It's a Goldilocks scenario," says Andrew Fruchter of the Space Telescope Science Institute in Baltimore, Maryland. "Only supernovae whose progenitor stars have lost some, but not too much, mass appear to be candidates for the formation of [long-duration] gamma-ray bursts."[21]

Fruchter is a member of a huge international team of astronomers known as GOSH (Gamma-ray Burst Optical Studies with HST) who study the environments of gamma-ray bursts. Their surveys show that long-duration gamma-ray bursts favor particular types of galaxies. "Most long-duration gamma-ray bursts are found in peculiar... low-metal, star-forming galaxies," he says, "which the Milky Way is not." The word "metals" in this case is astronomy-speak for elements heavier than hydrogen and helium that have been forged in stars via nuclear fusion. In the very early universe, the first generations of supermassive stars were hundreds of times more massive than the Sun, and there were a lot more of them at that epoch. But they didn't live very long—a few million years at most—and they blew up before they had a chance to produce an abundance of heavy metals. Hence, the galactic environments in which they resided were "metal-poor." Gradually, as star formation continued and the previously processed metals were recycled by newer generations of stars, the metal abundances of galaxies increased.

The low-metal preference may be why many long-duration gamma-ray bursts occur at distances of billions of light-years, a realm of the universe that appears as if it was less than 25 percent of its present age, when metals were fewer and star formation was running full tilt. There were more super-supernovae back then and consequently more long-duration gamma-ray bursts. Ten billion years later, though, more metals have been sown between the stars, and the heyday of supermassive stars is past—at least in our Galaxy. They can still occur in nearby galaxies, even in the neighboring Magellanic Clouds, in low-metal, star-forming hot spots, but from a distance of more than 100,000 light-years they pose no threat.[22]

That leaves us with the cause of short-duration gamma-ray bursts, those lasting between a few milliseconds and about two seconds. Here astronomers' certainty about the cosmic landscape recedes even further, and we enter the "here-be-dragons" realm of the map. Currently, the theory of choice is that a short burst might result from an interaction between two compact remnants of massive stars—either a merger between two neutron stars or from a black hole swallowing a neutron star. Either event would produce an outburst of gamma rays and X-rays that would be visible across the great gulfs of the galaxies. Astronomers are only now beginning to catch these objects in the act, or in the immediate afterglow of the burst. So far, most appear to pop up (or pop off) in old, star-spent elliptical galaxies, which further supports the theory that you need the dense by-products of old massive stars to ignite a short burst.[23] Moreover, elliptical galaxies contain many of the star-rich objects known as globular clusters, which would be the ideal breeding ground, since their crowded environment would provide the best chances for a stellar interaction.

Of course, the possibility exists that a maverick neutron star might be lurking nearby. Estimates from the current pulsar birth rate and the number of supernovae suggest that there could be as many as a billion neutron stars in the Galaxy,[24] the nearest lying at a distance of 260 light-years by some researchers' reckoning.[25] But nearby *binary* neutron stars capable of merging, and thereby touching off a deadly short-duration gamma-ray burst, are far less likely. So far, the nearest half-dozen neutron stars spotted are loners, and any binary system would likely have been detected by now as a bright X-ray source.

But there is yet another freakish class of stars that astronomers are now fairly certain exists in the Milky Way: neutron stars with intense magnetic fields that are called "magnetars." Although there is still much to be learned about them, there is little doubt that magnetars have affected Earth recently, and possibly destructively in the past. Their flash events resemble gamma-ray bursts but are different animals in two very singular ways. For one, they are strictly confined to the galactic plane, where all the star formation occurs (that's how astronomers know they are "of this Galaxy"). For another, unlike long-duration bursts and those touched off by merging neutron stars and black holes, magnetars do not elicit onetime events; they repeat themselves and are hence well plotted on the sky.

Magnetars are classified by how they are discovered. If they are detected when they flare, they are called "soft gamma-ray repeaters," or SGRs. If they are detected by their continuous X-ray emission and in a quiescent state, they are called "anomalous X-ray pulsars," or AXPs. Both, however, are the same type of beast and both are believed to be magnetars. As of this writing, there are sixteen known magnetars and candidates, six of which are SGRs and ten of which are AXPs.[26] Stay tuned for more.

Astronomers think magnetars are very young, probably less than 30,000 years old. Their precursors are very massive stars, greater than about 40 times the mass of the Sun, that have fast rotations and powerful magnetic fields, which are retained after collapsing to the neutron-star phase. Simply put, a magnetar's outburst occurs when torsional stresses within the star's magnetic field induces a "starquake" in its crust, releasing off-the-scale bursts of gamma-ray and X-ray energy.

The magnetar concept was first proposed in 1992 by Robert Duncan of the University of Texas at Austin and Christopher Thompson of the Canadian Institute for Theoretical Astrophysics in Toronto. At the time, it was greeted with no small measure of skepticism, but today astronomers are coming around to the idea that magnetars may be responsible for some record-holding gamma-ray flares. The first such event was detected in March 1979 when a wave of gamma radiation swept over ten spacecraft in the inner solar system. Two Soviet spacecraft, *Venera 11* and *Venera 12*, which three months earlier had dropped landing probes to the surface of Venus, were quietly orbiting the Sun when each registered a radiation count that went off-scale within a fraction of a millisecond. Eleven seconds later, detectors on NASA's solar probe *Helios 2* were similarly saturated. The Pioneer Venus Orbiter was next, followed seconds later by detectors on three US Department of Defense *Vela* satellites, the Soviet *Prognoz* 7 satellite,* and the Einstein X-ray Observatory. Finally, before exiting the inner solar system, it spiked detectors aboard the International Sun-Earth Explorer. The pulse of hard gamma radiation was 100 times as intense as any previous gamma-ray burst ever detected from beyond the solar system and was followed by a fainter glow of soft gamma rays and X-rays.

Over the ensuing years, astronomers continued to detect bursts coming from the same direction in the sky, as well as from many other directions, and hence the dialogue as to what caused them began.[27] Outbursts detected in 1979 and 1998 were powerful enough to zap orbiting satellites and ionize the upper atmosphere.

**Prognoz 7 was actually part of a French-Soviet collaboration to work with *Venera 11* and *12*'s complement of gamma-ray detectors as an interplanetary gamma-ray burst triangulation network.

Top: Drumlins are elongated hump-backed hillocks of material deposited by glaciers. The long axes of these features are parallel to the movement of the ice. This is an aerial photo of a drumlin field in central New York. Courtesy of Stan Celestian, Glendale Community College, Arizona.

Bottom: The glacier Mer de Glace reaching down to the floor of the Chamonix Valley in France, as seen from La Flégère in 1823 by artist Samuel Birmann. Notice the village Les Bois at the glacier's snout. Birmann's paintings of the Mer de Glace provide an important record of the glacier's growth in the first half of the nineteenth century. Kunstmuseum Basel, Kupferstichkabinett; photograph by Martin Buhler. Reprinted with permission.

Above: The Pencil Nebula, cataloged as NGC 2736, is the remnant of a star that exploded as a supernova 11,000 years ago. It is part of the huge Vela supernova remnant located in the southern constellation Vela. The banded structure is produced by shock waves (moving left to right) encountering regions of different gas density in the interstellar medium. The shock waves heat the gas to millions of degrees. The supernova, had it been observed, would have easily been visible to southern observers in daylight. Courtesy of NASA and the Hubble Heritage Team (STScI/AURA).

Top left: In this composite taken on December 2, 2003, the Sun can be seen at the center of the occulting disk used by the SOHO spacecraft to reveal the structure in the corona. At lower right, a large coronal mass ejection blows out a huge cloud of particles into space. Courtesy of SOHO (ESA/NASA).

Bottom Left: An artist's rendition of the gravity tractor, a robotic spacecraft that would use the mutual gravitational attraction between itself and an asteroid to slowly accelerate the asteroid in another direction. It would have to be deployed a decade or more prior to a potential impact. Courtesy of Daniel D. Durda (FIAAA) and the B612 Foundation.

Above: A night shot of several of the radio telescopes comprising the Allen Array in California, used in the Search for Extraterrestrial Intelligence. Courtesy of Seth Shostak/SETI Institute.

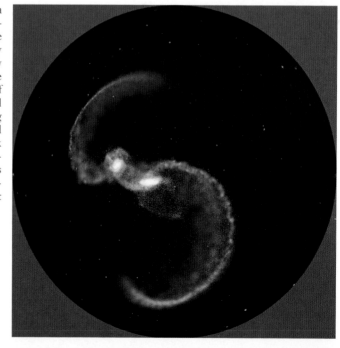

Right: A sample frame from a computer simulation of the upcoming collision between the Andromeda and Milky Way galaxies. At this stage, a few hundred million years after the first passage, the tidal forces of gravity have strongly distorted the galaxies' shapes, creating long plumes of material called "tidal tails." Courtesy of Frank Summers (Space Telescope Science Institute); Chris Mihos (Case Western Reserve University); and L. Hernquist (Harvard University).

In the 1998 case, the flare produced enhanced ionization at altitudes of between 30 and 90 kilometers (18 and 56 miles) that lasted for five minutes.[28] This from a magnetar located about 50,000 light-years from Earth.[29]

Then on December 27, 2004, the flash of a giant flare from a source known as SGR 1806-20 (the same object responsible for the 1979 flare) bounced off the Moon and, again, ionized the atmosphere and saturated satellite detectors. Though it lasted about two-tenths of a second, it was nonetheless 100 times more powerful than the flares of 1979 and 1998. Quantitatively, it amounted to an astounding energy release of 5×10^{46} ergs. The most powerful solar flares release about 10^{32} ergs of energy, or about 2 billion times the energy of a 1-megaton hydrogen bomb. This object, which is estimated to be some 32,000 light-years distant, released as much energy in a single sub-second burst as the Sun radiates in a quarter of a million years.[30] Moreover, the flare produced enhanced ionization of the lower daytime ionosphere down to about 20 kilometers (12 miles) for more than an hour.[31] Given the short time scientists have been monitoring these violent events, one wonders how regularly they have occurred in the past without our knowing about them, and how often they might crop up again in the future.

A typical magnetar has a field strength between 10^{14} and 10^{15} Gauss, which is some 10 trillion times stronger than a typical refrigerator magnet. (Earth's magnetic field is a wimpy 1 Gauss.) Think of a magnetar as a direct-current generator roughly the size of Washington, DC. The job of a generator is to convert mechanical energy into electricity, and that's what the magnetar does. But its magnetic field is the most powerful in the known universe—1,000 times that of most radio pulsars. The field is generated by the dynamo action resulting from the star's rapid rotation

—once every few seconds—and the roiling fluid of hot, ultra-dense neutrons circulating inside the star. The star's surface consists of a thin crystalline crust of neutrons through which the magnetic field moves. Occasionally, the violent churning of the fluid inside the star twists a patch of crust and, in so doing, causes the magnetic field lines to begin wrapping around one another like a braid. This, in turn, creates powerful shearing forces on the crust. When these become too great, the crust momentarily ruptures, releasing a fireball that emits a pulse of hard X-rays and soft gamma rays. Over the following minutes, the fireball fades until it finally vanishes altogether.[32]

These bursts do more than dissipate pent-up kinetic energy; they also slow the star's rotation rate, or so it is believed. The 1979 post-flare stage of SGR 1806-20 exhibited a train of 8-second pulses from the fading fireball, indicating a rotation rate of 8 seconds. When it was measured again in 1995 by the Rossi X-ray Timing Explorer, which was designed to be highly sensitive to fluctuations in X-ray output, the period had decreased to 7.47 seconds. Contrary to expectations, however, the December 2004 outburst did not show a further decline in the spin rate; it was still about 7.5 seconds.[33] Obviously, more observations on this aspect of magnetars are needed.

Armed with all this information, we can now assess the danger that a magnetar flare might pose to Earth. Clearly, if a magnetar strayed into the solar system, its magnetic field could very well scramble the atomic and molecular structure of the solar system. The magnetic field of a magnetar is powerful enough to split photons and distort the quantum vacuum. Fortunately, though, magnetars have very short lives. They are most active for their first 10,000 years. After that, the stars get colder and less magnetically active with age, and hence their X-ray brightness diminishes as well.

Although they are still likely to retain very strong magnetic fields long after 10,000 or 100,000 years, says Duncan, "the field is nearly frozen in by the cooling, superconducting core of the star, so magnetic energy dissipation occurs at much less significant rates."[34]

Duncan thinks it unlikely that a magnetar could be lurking near the Sun because it would have probably been detected as a bright X-ray source or associated with a young supernova remnant. But, he says, "because there seems to be considerable up-and-down variability over time in X-ray intensity, even in active, fairly young magnetars, we can't rule this out entirely." More dangerous, he adds, would be the supernova that created the magnetar in the first place. "Although a giant flare from a magnetar is 10,000 times brighter than a supernova for the initial few tenths of a second, as the December 2004 event was, its brief duration makes it much less damaging to the Earth's atmosphere and ozone layer than a supernova at the same distance. Thus, if a magnetar formed near the Earth, it is likely that the initial supernova that made it would be the most damaging thing to us. The flares that followed, within the first few tens of thousands of years, would just add insult to injury."

Geologic evidence indicates that stars have likely contributed to the direction that life has taken on Earth, and perhaps other habitable planets, through extinctions and genetic mutations. A supernova or gamma-ray burst may have been responsible, some argue, for Earth's most severe extinction event at the boundary between the Permian and Triassic periods 250 million years ago, as well as the second-largest mass extinction known, that of the Late Ordovician period, some 443 million years ago.[35] And why not? Reverse the Milky Way's stately rotation half a billion years, and the Sun wheels itself into a more volatile stellar environment. This has happened not once, but many times. Since its birth, the

Sun has made about 20 cycles around the Galaxy and passed many times through the spiral arms and different stellar neighborhoods. The effects these changes might have had on the solar system and on Earth will be discussed in more detail in the following chapter.

Today, we find ourselves in a calm part of the galactic whirlpool, and astronomers think the chances of being irradiated by a supernova, hypernova, or magnetar are at best remote. "It is likely that there exist many *old* magnetars within 3,000 light-years of the Sun," says Duncan, "but after a few times 10,000 years, these stars are unlikely to emit strong bursts that could threaten or affect the Earth." It may even be plausible, adds Duncan, that the Permo-Triassic extinction 250 million years ago "came from a supernova, perhaps forming a magnetar." (Though there are more prosaic, Earth-bound explanations for the extinction, including giant gassy belches from the ocean that upset the world's ecosystem.)

As I noted at the beginning of this chapter, stars do explode and they will continue to do so until the universe goes dark. But within a human life span, they are rare events. You may count yourself fortunate indeed if you ever see one with your own eyes (and preferably from a safe distance). The last supernova visible to the unaided eye occurred in February 1987, but it occurred in a neighboring galaxy, the Large Magellanic Cloud. It is still being studied, and its surrounding ring of shocked gas is being monitored as it evolves. The most recent visual supernova in the Milky Way occurred in 1604 CE. Thirty-two years before that, in 1572, Tycho Brahe saw the supernova that bears his name, and it was observed for about eighteen months.

Within the last decade, X-ray observations and Antarctic ice cores have revealed evidence for another supernova some two and a half centuries before Tycho's. That would put us closer to the estimate that two to three supernovae go off in the Galaxy each century. The age of a small supernova remnant in the southern constellation Vela—called Vela Jr.—correlates well with an excess of nitrate abundances detected in polar ice cores. Nitrates are salts that rain out of the sky when high-energy particles, such as gamma rays, react with nitrogen and oxygen in the atmosphere. When a sudden gust of high-energy particles collide with the Earth's atmosphere, they create a surplus of nitrates. It turns out that a well-delineated spike of nitrates was found in ice cores dating back to 1320 CE or thereabouts, and some astronomers interpret this feature as the geophysical calling card of a previous supernova.[36] The conclusions are still being debated, but the evidence is compelling. Similar nitrate abundances in polar ice cores date back to supernovae that occurred in the years 1604 (Kepler's Star), 1572 (Tycho's Star), and 1181 (the supernova widely witnessed by Chinese and Japanese observers).

Why the "1320" supernova was not observed is curious. Vela is a southern constellation visible from the southernmost realms of the Northern Hemisphere but visible anywhere in the Southern Hemisphere. It is also circumpolar—that is, it lies near the south celestial pole, hence it wheels into the sky somewhere every night. And further, it is only 650 light-years away. A supernova that close would have easily been seen in full daylight. Surely, such an apparition would have been noticed and recorded by the aboriginal folk in Australia or the Polynesians, all of whom were avid sky watchers. It would be an interesting exercise, indeed, to pore over the extant anthropological art of the fourteenth-century Southern Hemisphere cultures and try to forge a connection. But that is another subject for another day.

At any rate, the fact that no supernovae have been observed visually for at least four centuries would indicate that we are long overdue for another naked-eye supernova somewhere in our Galaxy, but, fortunately, nowhere near the Earth. The one wild card is Eta Carinae, which, at a distance of 7,600 light-years, is just near enough to give us pause. From what we can tell about the star's orientation and mass, its distance precludes any damaging effects on Earth when it blows. But there is still much that remains unknown about this singular sun. Will it shed enough mass to go out as a supernova or will it detonate as a hypernova? And when will it explode? Soon, or in a million years?

No one can say, except that, from all the observations, the star looks like it's about to let loose. When it finally blows, one thing it is sure to be is bright in the sky, particularly if it is anything like the type of supernova observed in another galaxy in 2006, designated SN 2006gy. Astronomers were astounded to discover that SN 2006gy was the most luminous and sustained supernova ever observed. The star that blew up as SN 2006gy would probably have looked a lot like Eta Carinae: it had an upper mass of more than 100 Suns, had not fully shed its hydrogen envelope, and was surrounded by a circumstellar nebula. In the decades before its death, it, too, would have been wildly unstable and volatile, experiencing explosive bouts of mass loss similar to the nineteenth-century eruptions of Eta Carinae.[37]

If Eta Carinae shares the fate of SN 2006gy, it could be 100 times more powerful than a typical supernova and appear almost as bright as the full Moon in the sky.[38] What would *that* do to us? As a standard supernova, it would irradiate the upper atmosphere, orbiting satellites, and unshielded astronauts, but if it's 100 times more powerful, the situation would become a bit more problematic. In short, we would be wise to keep a watchful eye on Eta Carinae and consider the possible consequences, particularly to satellites and spacecraft occupants.

I like to think that Eta Carinae has already gone supernova and that the light from the explosion is on its way. At its distance, we see the star as it appeared 7,600 years ago, give or take, before much of recorded human history was set down. So for all we know, it may have gone off thousands of years ago and the first waves from the detonation are approaching Earth's shore at this very moment. When the first waves break, all faces will turn skyward as the star brightens, then, unbelievably, brightens even more. It will be a transcendent sight, like a bright comet or a flashing bolide, but far more rare. After it fades, astronomers will turn their telescopes and spacecraft upon the glorious aftermath, analyze and map the tangle of expanding and ionized shock waves, search for a spinning remnant, and assay the elemental content of its dissipating clouds. Such a celestial spectacle, one not seen for centuries, will no doubt rattle our cosmic cage. Do you think anyone will pause to reflect upon the fact that the collimated force of the star's blast was directed just slightly away from the galactic plane and roughly in the direction of a spiral arm containing millions of stars and who knows how many planets, and wonder what, or who, may have been given to the fire?

CHAPTER 6

THERE GOES
THE NEIGHBORHOOD

The galaxy is a flung thing, loose in the night, and our solar system is one of many dotted campfires ringed with tossed rocks. What shall we sing?
—Annie Dillard, *Teaching a Stone to Talk*

On November 17, 1955, a group of scientists at the Naval Research Laboratory launched a rocket from White Sands Proving Ground in New Mexico. Their purpose was to survey the night sky for far-ultraviolet radiation. At an altitude of 104 kilometers (65 miles), the instrument's collimated photon counters detected a diffuse fluorescent glow of "Lyman-alpha" radiation, a strong spectral signature for swarms of neutral hydrogen atoms.[1] In interstellar space, hydrogen atoms are cold and hence their electrons remain in a neutral, or ground, state. However, when they drift into the vicinity of the Sun, photons generated in the solar atmosphere excite the hydrogen atoms out of their neutral state so that their electrons leap to their first excited energy level. When the electrons decay back to their ground state, energy is released in the form of Lyman-alpha photons.

That kind of radiation, called "airglow," was expected to be found in Earth's upper atmosphere. But the Lyman-alpha radiation that the navy scientists detected that day was different. This glow seemed to come from beyond the atmosphere. It was so bright, in fact, that celestial sources of Lyman-alpha light could not be detected through it. Though the interpretation of the observations was debated for a few years to come, astronomers finally realized they had detected something remarkable: far-ultraviolet radiation from the interstellar medium. Interstellar space was humming with energy.

It was a modest rocket experiment but an important one with results that shouldn't have surprised astronomers. By the mid-1950s, the nature of interstellar space, which had been a topic of interest for centuries (dating back to the early seventeenth century), was rapidly giving up its secrets. Throughout the late nineteenth and early twentieth centuries, a series of photographic, spectroscopic, photometric, and radio observations began to "fill up" the interstellar medium with something more substantive than light-bearing ether.

What amounted to "first contact" with the interstellar medium occurred in 1904 at Potsdam Observatory when German astronomer Johannes Hartmann studied the spectrum of the star Mintaka—the westernmost star of the three that form Orion's Belt. Mintaka is actually a binary star whose closely orbiting neighbor causes it to wobble or oscillate in position (astronomers back then called them "variable velocity stars"). Hartmann found that all of the star's spectral lines shifted in accordance with its oscillations—save those of calcium, which remained in their terrestrial locations throughout the oscillation period. This meant that calcium atoms were "stationary" with respect to the Earth. Hartmann reasoned that calcium atoms must be sprinkled in space somewhere between

Earth and Mintaka.[2] Fifteen years later, another astronomer, Mary Lea Heger of Lick Observatory, made second contact with the interstellar medium when she noted "stationary" sodium absorption lines in the spectra of two stars, Beta Scorpii (also a close binary star) and, once again, Mintaka.[3]

In 1919, American astronomer Edward Emerson Barnard published an atlas of remarkable photographs of "dark markings" among the stars in the Milky Way that revealed that these were, in fact, regions of obscuring interstellar matter (though Barnard himself considered some of them "vacancies").[4] A few years later, German astronomer Max Wolf provided convincing evidence that confirmed these dark areas were obscuring interstellar clouds.[5] The nature of these clouds remained a mystery for many years until observations revealed them to be composed of solid micron-size dust particles.[6] This dust was later associated with the reddening of starlight.

Observations and discoveries continued apace. By 1951 the detection of radio emission from cold interstellar hydrogen became step one toward describing the temperature, density, and motion of interstellar matter.[7] The identification of the more energetic Lyman-alpha radiation, mentioned above, was a solid step two.

The most recent observations (too numerous to list here) have taken us many steps beyond. In sum, their findings have done far more than quantify the properties and content of the spaces between the stars—they have woven the stars themselves, and their planets if any, into the fabric of the Galaxy. In this new model, the components of the Milky Way, though separated by great distances, are nonetheless interrelated, even though different regions may vary quite a bit in temperature, density, and state of motion. An event happening "here," a supernova, for example, can enhance or impede star formation "there." Stars orbiting on one side of the Galaxy can be perturbed by encounters with stars or interstellar clouds on the other side. In short, the Milky Way is more than just a place where we happen to find our solar system and ourselves: it is our cosmic world, and just as events occurring in the deep ocean can affect events on land, the regions between the stars can affect what happens to the Sun and the Earth. Though interstellar space may be vast, the reach of the stars is long.

From our vantage point on Earth, we can't see the Milky Way as the vast wheel of stars that it is. We see it only as this band of gray, grainy light crossing from north to south—that is, if we see it at all. From most large cities aglow with their own brash illumination, the Milky Way is hardly noticeable. You have to make some effort to see it, such as drive out into the country, venture into the mountains or desert, take a sea cruise, or suffer a major power outage. In other words, you won't find the Milky Way from the environs of a shopping mall.

The best time of year to see the Milky Way is in midsummer, when it is highest in the evening sky as darkness falls. (There is also a winter component that is also lovely, though it doesn't quite have the optical presence the summer Milky Way has.) But by far, there's no better location to appreciate the *whole* of the Milky Way than the Southern Hemisphere. I was fortunate enough to have been born in South Texas where the Milky Way's central region, located in the direction of the constellation Sagittarius, rises to a respectable altitude of about 35°, a little more than a third of the way up from the southern horizon. It's a lovely vantage point, to be sure, especially when compared with that of more northern locales, where the great galactic center might barely edge above the tree line. Still, even from the southern realms of the Northern Hemisphere, there is so much more of our Galaxy that lies below the horizon.

I finally got my chance to see just how much more in August 2007, when my wife and I went to Sydney, Australia, to visit my good friend Gary, with whom I had shared an avid interest in amateur astronomy way back in high school. Gary was eager to give us a tour of his "getaway" retreat on Australia's Pacific coast, a village called Hawks Nest in New South Wales. But what he really wanted to do was show us the Milky Way. The last time he and I had seen it together was as teenagers from a dark and spooky cornfield west of our hometown, Corpus Christi. But that was from a latitude of 27.5° *north*.

Of course, even without the Milky Way, the location was delightful. Hawks Nest lies on a peninsula that juts into the blue-green waters of 7-mile-long Nelson Bay. On the north side—the Hawks Nest side—lies a beautiful arc of white beach called Jimmy's Beach that extends east to the Pacific Ocean and a striking tree-covered promontory called Yaccaba Head. On the south side of Nelson Bay lies Port Stephens, a popular resort area and consequently more crowded with tourists. Hawks Nest, pre-ferred by the Aussies, offers a slower pace and greater seclusion.

At twilight we walked the few blocks from Gary's house to Jimmy's Beach, which we had to ourselves. We lay out a blanket, opened a few bottles of beer, and waited for the stars to come out. And did they come out! As I said, from South Texas the center of the Milky Way gets about a third of the way up in the sky, but from latitude 32° *south*, it floats high overhead, a distinct bulbous glow mottled by channels of dark interstellar dust. From either side of the bulb extend both segments of the Milky Way, one of which billows down to the southern horizon, the other to the northern horizon and into the Pacific Ocean. To the left, or south-east, of the main band, I could also make out the patchy Small Magellanic Cloud and, a bit lower, the L-shaped Large Magellanic Cloud, both satellite galaxies of the Milky Way. It was my first time to see them.

Suffice it to say that the vista was a stunner (or, as they say down under, I was gobsmacked), and my friend Gary was amused as I went on and on about the view. With such a celestial panoply laid out before you, it doesn't take a lot of imagination to envision the great arms of the Galaxy sweeping out of the north, passing broadly in front of the central bulb, and trailing southward off in the distance around the far side of the Galaxy. As a science writer, I have spent most of my life intellectualizing the Milky Way both in print and vocally as a multiarmed spiral galaxy, just like many other multiarmed disk galaxies in the universe. But from Australia, the Milky Way is something more visceral, entirely other than an abstraction. It isn't just a band of light in the night sky: it is a *wheel*. And, moreover, you realize we are *in* the wheel—and that the wheel, and everything in it, is moving. Such was my epiphany at Hawks Nest!

I came back from our trip thinking that the Milky Way truly is a "flung thing," as Annie Dillard describes it, and the Sun along with it. But this reality makes our existence a bit more dicey. Over its 4.5-billion-year lifetime, the Sun has crossed the spiral arms of our Galaxy at least 17 times and encountered massive interstellar clouds as many as 50 to 60 times.[8] Evidence suggests these various environments can potentially cause a chain reaction of biospheric events. For example, the enhanced rate of star formation and supernovae in various regions of the spiral arms have likely created periods of more intense exposure to cosmic rays on Earth millions of years ago. This, say some scientists, could have affected the atmospheric ionization rate and the formation of charged aerosols that promote cloud condensation at low altitudes. Low cloud cover is very effective at increasing surface cooling by reflecting sunlight back into space. A prolonged period of such conditions could precipitate ice ages and, perhaps, mass extinctions.

Even now, 98 percent of the heliosphere—the volume of space filled by the solar wind—is made up of interstellar matter, harmless bits of atomic particles and dust that constantly pass through interplanetary space.[9] But life on Earth would be very different if we were located somewhere else, such as in or near a hot star-forming region. As it is, the heliosphere and atmosphere shield us from the raw and possibly detrimental environment of the interstellar medium. But over the history of the Sun, that hasn't always been so. Occasionally it's been as if a window has opened, creating a decided draft.

A century of observations have allowed astronomers to assay the galactic environment between the stars out to thousands of light-years. They have found that it's a clumpy region filled with the drifting filaments of dead stars, dense knots of cold dust, supernova shock fronts, and clouds of hydrogen gas. For the last several million years, the Sun inhabited an extremely vacuous region of the Milky Way, but it is now entering a denser region of the interstellar medium called the Local Interstellar Cloud. There is a "wind" here of low-density interstellar gas blowing over the solar system from the direction of the galactic center, which is driven by supernovae long gone. Radio telescopes have detected Giant Molecular Clouds, impenetrable regions of inter-stellar gas composed largely of molecular hydrogen, which also harbor volatile protostars. Infrared surveys and spectroscopic observations indicate that within a few thousand light-years, the interstellar medium exhibits a broad range of temperatures, den-sities, and compositions. And given that the Sun moves through the Galaxy as well, it is obvious that, over its lifetime, it has passed through a variety of galactic environments. Fortunately for us, we still abide in a relatively quiet region of the Milky Way—but that won't always be the case.[10]

We often think of the night sky as a frozen tableau, static and unchanging—and when you consider that the stars tonight are essentially the same stars Galileo observed four centuries ago, that might seem a reasonable assumption. But, as the Greek philosopher Heraclitus put it more than 2,500 years ago: "All is flux, nothing stays still." Indeed, everything in the universe is in motion, and the Sun is no exception. Its orbital velocity is about 225 kilometers (140 miles) per second. Per year, that adds up to over 7 billion kilometers (4.3 billion miles) of interstellar space, or slightly less than the distance between the Sun and Pluto when it is at aphelion. Over a human lifetime (let's say 72 years), the Sun traverses more than 510 billion kilometers (316 billion miles), about 0.054 of a light-year. It takes more than 1,300 years to break the 1-light-year threshold and about 225 million years to complete one circuit about the Galaxy.

During a galactic year, stars come and go, constellations change shape, and the view toward the inner and outer Milky Way morphs as curtains of dust first obscure then reveal distant star clouds. On Earth, except for some long-enduring kinds of bacteria (cyanobacteria), algae (stromatolites), and brachiopods, life comes and goes. The dinosaurs appear and disappear within a period of 135 million years, which amounts to slightly over 50 percent of the orbital period of the Sun—a good long run. History dating back to stone-wielding hominids represents about 1 percent of that period; back to the first North Americans, less than 0.01 percent. For more modern periods, the intervals are too trivial to mention.

So compared to the brief flicker of a human lifetime, or even

1,000 human lifetimes, the Sun's motion, though it *seems* glacial, is an illusion created by objects that are incredibly far away moving through an incomprehensibly vast space. It takes time—millions of years—for the smallest changes to become noticeable (by eye) to us poor short-lived humans.

Even if the Sun's pace were more lively, we are limited in our observations of the Galaxy by another factor: interstellar dust. The majority of all the stars we can see in the sky on any night can be mostly contained within a radius of about 7,000 light-years. Such a radius reaches the dusty star clouds of the winter Milky Way in the northern sky and those in the summer Milky Way in the southern sky. The beautiful Double Cluster in Perseus, easily visible on a dark, clear winter's night, is about 7,000 light-years away. In the opposite direction lie many star-forming nebulae apparent to the unaided eye, including the Lagoon and Trifid nebulae in Sagittarius (about 5,000 light-years away) and the Eagle Nebula (7,000 light-years away).

This chunk of galactic real estate, which makes up a very small part of the Galaxy, is, in terms of visual wavelengths, a line-of-sight zone constrained mainly by how much intervening interstellar dust we have to peer through. In some directions, we can see much farther than 7,000 light-years, particularly if we grab a telescope and look toward the Galaxy's population of globular clusters. These great spheres of stars are situated in the galactic halo, well above the obscuring dust in the plane. Their distances are typically on the order of tens of thousands of light-years. When our line of sight is perpendicular to the disk of our Galaxy, where dust extinction is negligible, the most distant stars we can see with our eyes are, on average, only 400 to 500 light-years away. But that's because stars tend not to stray far above the gravitational confines of the disk, unless their vertical velocities are extreme. On the other hand, if we aim our telescope in that direction, we see other galaxies near and far—hundreds of thousands to hundreds of millions of light-years distant. The nearest large array of galaxies, the Virgo Cluster, is scattered throughout the spring sky. As we will later see in this chapter, this sprawling cluster, though it lies over 48 million light-years away, might well have had a detrimental effect on our humble solar neighborhood millions of years ago, and could again.

It would seem, then, that other than providing us with a big picture of our tiny corner of the universe, the visual feast we see in the night sky says little about our solar neighborhood. It's like trying to discern our location and environment in a forest by surveying distant mountains and rivers. We can do better, however, if we restrict ourselves to making fine positional measurements of nearby stars by using telescopes that can detect slight sub-arcsecond changes in the motions of stars with respect to more distant background stars. Only this way can we apprehend where our Sun has been in the Milky Way and where it's going—Earth along with it.

Where are we going? There is no destination, per se. Where are people going on a carousel or a Ferris wheel? Round and round and up and down. On the other hand, we have a pretty good idea of our speed and direction. Relative to the stars in our part of the Galaxy, the Sun is moving at a velocity of about 19.5 kilometers (12 miles) per second in the direction of the constellation Hercules, southwest of the bright star Vega and just north of the billowy clouds of the summer Milky Way. (Let us remember this for later.) This point in the sky is referred to as the solar apex. It is determined by measuring the spectral Doppler shifts of stars moving toward or away from the Sun, as well as their motions measured transversely on the sky (in terms of radial velocity and proper motion, respectively). Except for stars nearest the Sun (such as Sirius, Alpha Centauri, and Barnard's Star), these motions are typically measured in milli-arcsecond units. In the direction opposite the apex lies the Sun's "antapex," which is in the constellation Columba in the southern sky.

Of course, it's a little more complicated than that. In fact, it's a *lot* more complicated than that. There are a host of factors to consider when describing the Sun's journey through the Galaxy, including its motion relative to neighboring stars, how its noncircular orbit affects its future path, and how encounters with interstellar clouds or other stars might perturb the orbit. We will touch on some of these issues, but since we are mainly concerned here with how the Sun's motion through the Galaxy might affect Earth, a technical tangent along these lines is not only unnecessary but would divert us from our purpose. A broad view and a few specifics will more than suffice.

Astronomers view the Sun's trajectory through the Galaxy, as well as the trajectories of all nearby stars, relative to an imposed frame of reference called the "local standard of rest." Basically, this is an imaginary volume of space about 300 light-years in diameter centered on the Sun that moves in a perfectly circular orbit around the Galaxy (clockwise, as viewed from north of the galactic plane). It might help to think of the stars as being contained in a "globe" that circles the Galaxy. As such, all the stars within the confines of the globe are "at rest" and have space velocities that average to zero.

However, as we found out a few paragraphs ago, we know that the Sun moves relative to the local standard of rest at 19.5 kilometers (12 miles) per second toward the solar apex. We also know that the other stars must be moving as well in directions that deviate from the local standard of rest. And, indeed, when we measure the radial velocities (line-of-sight motion) and proper motions (transverse motion) of the stars around us, that is exactly what we find. The deviations, or "peculiar" motions, are the result of their actual motion through space. Their motions also reflect the Sun's motion toward the solar apex, depending on where they are located with respect to the solar apex. Stars located in the direction of the solar apex will exhibit diverging radial velocities, while those near the antapex will exhibit converging radial velocities. Stars located 90 degrees from the apex and antapex (i.e., adjacent to the Sun's direction of travel) will tend to exhibit the largest proper motions toward the antapex. Although not an exact analog, think of the "warp" effect of the starship *Enterprise* in *Star Trek*.

To pin down the Sun's orbit (as well as those of other stars), we need to know its three velocity components with respect to the galactic center: how fast it's moving toward (or away from) the galactic center; its transverse (or forward) speed through the plane; and its vertical velocity in the plane. All three must be factored in to derive the Sun's true motion. After years of calculations and still no definite consensus, astronomers generally maintain that the Sun is moving toward the galactic center at a velocity of 10.4 kilometers (6.4 miles) per second. It is moving through the plane at 14.8 kilometers (9 miles) per second. And the vertical motion is 7.3 kilometers (4.5 miles) per second. Taken all together this yields the solar apex motion of 19.5 kilometers (12 miles) per second.[11]

This estimate was originally based on bright nearby stars irrespective of spectral class, something that could bias the results. Interestingly, however, when using a larger unbiased sample of stars from the *Hipparcos Catalogue*, known for its 1-milli-arcsecond-tolerant star positions, and omitting the brightest nearby stars, these values are 10.0, 5.3, and 7.2 kilometers (6.2, 3.3, and 4.4 miles) per second, respectively, yielding an apex velocity of 13.4

kilometers (8.3 miles) per second.[12] Given that these spatial velocities are modest, the Sun's orbit can be described as nearly circular, just like Earth's orbit around the Sun. Stars with similar values cling more or less to the galactic plane. In general, the larger the spatial velocity values are, the more a star's orbit deviates from the local standard of rest; its orbit is more eccentric and can take it farther away from the galactic plane.

With all these various motions spinning in our heads, let's employ a little imagination to take a look at where the Sun is in the Milky Way. If we could float above the galactic plane, say 3,000 to 4,000 light-years, we wouldn't be able to see our Sun even with the aid of a telescope. But we could just make out its location, near a dusky pocket adjacent to a tangled juncture of starry arcs made up of the stubby Orion Spur, to which we more or less belong, and the grand, sweeping star-capped ridges of the Sagittarius, Crux, and Carina arms, which lie between us and Galactic Central. Within a 3,000-light-year radius we would see tight knots of stars here and there, pink and blue wisps of gas, and an abundance of dark clouds of dust obscuring more distant stars beyond. Zooming in to within 20 light-years, we would finally be able to see our yellow star surrounded by a sparse cluster of fifty or sixty stars, a half dozen of which would be quite bright indeed. The vast majority, however, would appear dim and red, like embers in ash. The plane of the solar system, if we could make it out, would be a minuscule disk inclined nearly perpendicular to the galactic plane, like a mainsail bent from vertical in the wind.

This region, quiet as it seems, has, in fact, seen a lot of action in the last several million years, mostly in the form of supernovae. Trailing the Sun by about 1,000 light-years is the center of the Gum Nebula, a supernova remnant that covers about 40 degrees of sky (or about the width of 80 full Moons) in the Southern

Hemisphere. It's so large and tenuous that it's too faint to be seen with the naked eye. The supernova that created this nebular complex is estimated to have occurred about 1 million years ago. Also located in the direction of the Gum Nebula are several other supernovae remnants. One, located about 815 light-years away in Vela, may have occurred about 11,000 years ago.[13]

There are also a handful of stars in the Sun's vicinity that have "passed on" after puffing off their outer shells. Commonly referred to as planetary nebulae for their disklike appearance in small telescopes, these objects provide us with a vivid glimpse into the future of our Sun in another five billion years, after it has consumed most of its nuclear fuel. The nearest planetary nebula, which lies in the direction of the constellation Aquarius, is called the Helix Nebula.

Now let's concentrate on our solar environs by tightening our radius so that it's only 1,000 light-years in extent, and apply radio, infrared, and ultraviolet observations to our surroundings. We see that our Sun is presently located just outside of a hot (about 1 million degrees Celsius or 1.8 million degrees Fahrenheit), ionized region of interstellar space some 600 light-years across (and larger in places) known as the Local Bubble. In truth, the "bubble" is a cavity carved out of the distribution of clumpier bits of interstellar gas. The proximity of the Sun to the Local Bubble has had a far-reaching effect before on the heliosphere, a "pocket" of ionized gas blown by the solar wind into the interstellar medium.* During the time that the Sun was embedded in the rarefied and fully ionized plasma of the Local Bubble, there were no neutral interstellar atoms in the heliosphere. In fact, in such a low-density

---

*Unless otherwise specified, and to minimize confusion and endless parenthetical inserts, I use the term *heliosphere* to include the termination shock (where the solar wind goes subsonic), the heliosheath (beyond the termination shock, where the solar wind slows even more and compresses), and the heliopause (where the solar wind's pressure is equal with that of the interstellar medium).

environment, the heliosphere would have expanded beyond its current dimensions, producing a thicker heliosheath (the region where the solar wind begins to mix with the interstellar medium). This, in turn, would have shielded the inner planets from galactic cosmic rays, which, as their name implies, come from outside the solar system but are, in fact, particles. Today, the solar system is in a slightly more dense environment, the heliosphere is smaller, and the interstellar neutral particles that approach the Sun can be ionized and accelerated by the solar wind out to the heliosphere's "termination shock"—where the solar wind becomes subsonic. These particles then rain back into the inner solar system as "anomalous cosmic rays," so-called because their energies are too low to escape the heliosphere compared to the impinging galactic cosmic rays.

More important, a less robust heliosphere allows galactic cosmic rays to reach Earth, where they collide with molecules in the atmosphere, creating elementary particles called muons. About 10,000 muons strike every square meter of Earth's surface per minute. You might have seen them before, making wispy tracks in a cloud chamber. It is generally agreed that these particles are harmless to humans and animals, although some scientists say there may be exceptions to the rule. They point out that during the millions of years in which the solar system was immersed in this benign region of the Galaxy, *Homo sapiens* evolved and civilization emerged.[14] But this also implies that there must have been periods of millions of years when the Sun did not reside in such a safe harbor. Of course, this is another of those "cosmic coincidences" that, depending on your viewpoint, is either by design or by accident. Either way, it demonstrates how life can arise and abide in an otherwise hostile universe.

The most sensitive contact point between the solar system and the interstellar medium is the heliosphere and its components. Think of the heliosphere as a candle flame, and the various densities of interstellar material it encounters as puffs of wind that cause the flame to flicker. In the Local Bubble, the flame is steady and large, but outside this quiet environment, the flame is subject to winds of whatever interstellar gas or stellar environment it encounters on timescales between 1 million and 5 million years. The heliosphere has evidently been modified in the past by various interstellar states, and it will no doubt be again. For us, the question is what effect, if any, will these changes have on Earth?

The most recent heliospheric contraction due to the interstellar medium occurred sometime between 1,000 and 140,000 years ago, when the Sun exited the Local Bubble and entered a warm, tenuous region of low-density interstellar "cloudlets" flowing at a velocity of 26 kilometers (16 miles) per second from the direction of the Scorpius-Centaurus Association. This region is a hot spot for producing supernovae and shells of outflowing gas. At a distance of about 1,300 light-years, the Scorpius-Centaurus Association is the nearest star-forming region to the outskirts of the Local Bubble.[15] Tracing back the trajectories of its stars and other nearby stellar associations for the past twenty million years, astronomers have been able to estimate that at least one or more supernovae occur in the Sun's vicinity every million years or so.[16] Clearly, during its journey around the Galaxy, the Sun has been bombarded by high-energy photons and particles from nearby supernovae. It is thought that one or more supernovae may have created the Local Bubble's rarefied interior.

The interstellar medium immediately surrounding the Sun's heliosphere, called the Local Interstellar Cloud, is one of the best examples, per volume, of nothingness in the known Galaxy. The density of a cold cloud of molecular hydrogen, the birthplace of massive stars, is about 1,000 atoms per cubic centimeter. Diffuse nebulae have average densities of between 10 and 100 atoms per cubic centimeter. But the average density in nearby local interstellar gas is less than 0.1 atom per cubic centimeter. By comparison, a cubic centimeter of sea-level atmosphere would have to be spread out over 163 light-years to have the same spatial density as nearby interstellar gas.[17]

But the Sun may, in a few tens of thousands of years, encounter significantly denser clouds, perhaps outflow from the Scorpius-Centaurus Association or from some other part of the local region. These clouds might not have the size to be immediately noticed in surveys but could be large enough to cause big problems. There is at least one reason not to dismiss this notion out of hand. New interstellar clouds are being discovered all the time. For years now, astronomers have mapped the presence of galactic fluff in and around the galactic plane. Some members of a class of "high-velocity clouds" discovered in 1963 have been clocked moving at over 100 kilometers (62 miles) per second—well above the velocity expected from just the Galaxy's rotation alone. Most, however, lie at great distances, between 5,000 and 14,000 light-years away in the galactic halo.

A little closer to home, radio observations made in 1983 were matched up with a diffuse cloud first observed on the Palomar Observatory Sky Survey plates. Known as the Draco Molecular Cloud, it extends over 100 light-years in diameter and lies at a distance of 3,200 light-years. Moreover, it is approaching Earth at 21 kilometers (13 miles) per second. What makes this cloud so unusual, and potentially dangerous to any habitable planet it encounters, is that it is swept back into billions of dense gaseous plumes, ranging in size between 2 and 20 light-years, prompting a number of astronomers to refer to it as a "cometary cloud" (though it is not cometary in nature).[18] It is thought that the Draco Cloud may be gas plunging into our Galaxy from either an extragalactic or a galactic source and that it is unlikely to survive passage through the dense interstellar medium of the galactic plane.[19] If our Sun were to collide with such a cloud, it would certainly have unpleasant consequences for us. Fortunately, by the time it collides with our galaxy, about 44 million years from now, the Sun's own motion, as well as a transverse component of the cloud's motion, will steer the Earth clear of danger.

But to reiterate my point, the interstellar medium is not homogeneous, and not all clouds and cloudlets in the solar neighborhood have been accounted for—neither has every shock wave, temperature change, or ionization front in the interstellar medium. The Sun is always moving. In 3 million years it travels over 160 light-years, and the interval between cloud encounters is thought to be less than 80,000 years.[20] In periods as short as several thousand years, the Sun's heliosphere will traverse through different interstellar environments. An unknown cloudlet of higher density or a radiative shock from a past supernova could surprise us, and relatively quickly. The latter phenomenon, given a velocity in the 100-kilometer-per-second range, could cross the heliosphere in a matter of 100 years. Such regions could be common in the gulfs of the interstellar medium.[21]

Currently, the termination shock of the heliosphere lies well beyond the planets at a distance of nearly 100 AU (1 AU, or "astronomical unit," being equivalent to the mean distance between the Earth and the Sun). If the Sun encounters denser interstellar cloudlets, however, the heliosphere will contract toward the Sun, maybe a little, maybe a lot. A high-density cloudlet, say 1,000 atoms per cubic centimeter, could conceivably shrink the heliosphere to within the orbit of Saturn.[22] If that occurred, the environment around Earth would be saturated with interstellar gas. Moreover, if the dust in the cloud was great enough, it could attenuate sunlight for tens of thousands of years.

Fortunately, an encounter with a high-density cloud probably won't occur for another several million years at least. What is more likely, however, is an encounter with an uncharted region of slightly greater density than the one we presently occupy. Even that, however, could have serious consequences for Earth's climate. Recall that the heliosphere, Earth's magnetic field, and Earth's atmosphere collectively modulate how many cosmic rays reach Earth's surface. If the heliosphere were to be compressed by a denser interstellar environment, it would mean a significant increase in the flux of energetic cosmic rays at the top of the Earth's magnetosphere, perhaps 10 to 100 times the present rate and at energies between 10 and 100 mega-electron volts.[23] As described earlier, cosmic rays ionize particles in the upper atmosphere, creating showers of secondary particles that lead to the production of muons in the lower atmosphere. Beginning in the 1990s, research by some scientists indicated that muons and other ions may serve as low-altitude cloud nucleation sites.[24] Why low altitude, specifically? Because galactic cosmic rays have the energy to penetrate more often to an altitude of at least 2 kilometers (1.2 miles), while intermediate- and low-energy cosmic rays are more often foiled from reaching that level by the Earth's magnetic field and atmosphere. The point these scientists were making was that an increase in cosmic rays would mean an increase in low cloud cover that would, in turn, significantly increase our planet's "albedo," or reflectivity. Over a prolonged period, as sunlight continually reflected back into space, global temperatures would fall, possibly creating another ice age, or so goes the theory.

And for several years, that's all it was: a theory, but based on a little fact. As our earlier detailed description of the Little Ice Age showed, the solar magnetic field during periods of high solar activity acts as a natural force field by shielding the inner solar system from the incoming cosmic rays. Hence, a less-active Sun usually means more cosmic rays reaching the Earth's surface. But how does a slight decline in solar irradiance (0.1 percent) and a greater flux of cosmic rays translate into a significant downturn in global temperature? A hint of an answer came to light in 2000, when a correlation between cosmic-ray flux and cloud cover was found over the duration of a single solar cycle.[25] Now an answer could be proposed: a prolonged low-energy, low-magnetic Sun allowed a steady rain of cosmic rays to infiltrate Earth's lower altitudes. Over the decades and centuries, they fostered an entrenched period of low cloud cover, which resulted in a higher albedo and cooler temperatures during much of the Little Ice Age.

Okay, fine, but this is a wobbly solution at best. Are there any facts to support this hypothesis? Not yet, but there is yet another grand assumption that seems more than reasonable to many scientists. In our Galaxy, the densest interstellar environments that could severely compress the heliosheath and expose the inner solar system to a raw interstellar medium would be found in the spiral arms. In the spiral arms you would also find the hottest, most massive stars in the Galaxy, the kind that tend to go supernova in just a few tens of millions of years. Since galactic cosmic rays are thought to be generated in supernova shock waves (and other energetic bangs like gamma-ray bursts), the Sun's (and, by association, the Earth's) passage through a spiral arm would mean greater cosmic-ray exposure for the solar system in general. In theory, then, the cosmic ray–induced cloud cover would result in cooler Earth temperatures during the time in which the Sun traversed the spiral arm, for hundreds of thousands of years at least.

Based on the Sun's current position and velocity, as determined from the distances and motions of nearby stars, astronomers have been able to track the path of the solar system through the Galaxy for the past 500 million years. They find that the Sun could have crossed four times through segments of the spiral arms and, moreover, that if and when these passages occurred, they would correspond to major ice ages and, perhaps, mass extinctions.[26]

Corroborating this finding is another study in which the cosmic-ray exposure ages of iron meteorites, as well as isotopic evidence from ice cores, point to a periodicity in cosmic-ray production of 143 million years (plus or minus 10 million).[27] If cosmic-ray production varies with this kind of period—and, in the interest of full disclosure, not all scientists think that it does[28]—it would link changes in the solar neighborhood directly to climate and the evolution of life on this planet. Such an idea was first considered by American astronomer Harlow Shapley way back in 1921. Shapley speculated that an 80 percent change in solar radiation due to the irregularity of interstellar clouds "would completely desiccate or congeal the surface of the Earth."[29] Let's hope it never comes to that!

All of these hip-bone-connected-to-the-knee-bone extrapolations, fascinating though they may be, are still lacking an answer to a fundamental question: can cosmic-ray exposure create clouds? It's a subject that has been debated and argued and harrumphed over for years, but there may now be an answer that at least some scientists may feel comfortable taking to the bank.

Henrik Svensmark, director of the Center for Sun-Climate Research at the Danish National Space Center in Copenhagen,

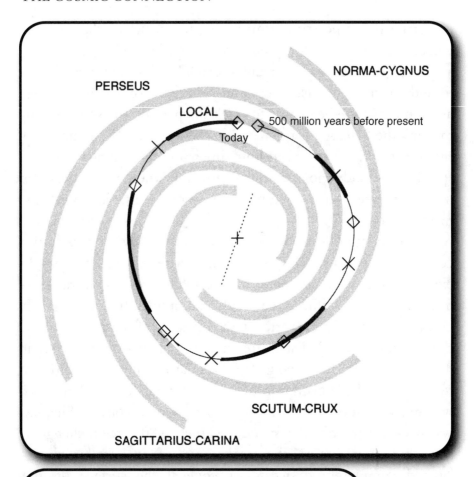

Figure 12. Tracing the Sun's clockwise motion through the Galaxy over the last 500 million years shows that it has traversed four spiral arms (gray bands) at times that correspond with ice ages. In this schematic, the thick solid lines correspond to ice age times and the crosses indicate times of mass extinctions. Diamonds mark intervals of 100 million years. The plus sign indicates the center of the Galaxy, and the dotted line the location of its central bar. Illustration courtesy of D. R. Gies and J. W. Helsel, Georgia State University.

thinks cosmic rays can answer for climate change, even (albeit controversially) part of global warming during the twentieth century. Svensmark and his team built an experiment called "Sky" in the basement of the DNSC that was designed to trace the chemical action of cosmic rays in a large plastic reaction chamber. The chamber contained purified air and the trace gases that one would typically find in unpolluted air over the ocean. To supply a source of cosmic rays, ultraviolet lamps were employed, with gamma rays reserved when they wanted to test higher amounts of ionization. In effect, the scientists provided the cosmic rays, and their reaction chamber served as a scaled-down model of Earth's atmosphere. After several years of running the experiment, their data showed that electrons created by the ultraviolet light significantly accelerated the formation of aerosols about which cloud droplets coalesce, filling their reaction chamber with microscopic droplets. When they zapped it with gamma rays, droplet production increased proportionately. Said Svensmark in a DNSC press release: "The odds are 10,000 to 1 against this unexpected link between cosmic rays and the variable state of the biosphere being just a coincidence, and it offers a new perspective on the connection between the evolution of the Milky Way and the entire history of life."[30]

It needs to be pointed out here that few climatologists accept Svensmark's claims that cosmic rays, or the lack of them, can affect climate in the short-term—nor can solar variations on decadal timescales. At least one recent study shows that global warming since 1985 is due to neither an increase in solar radiation nor a decrease in the flux of galactic cosmic rays.[31] On longer timescales, however, the Sun and cosmic rays may be linked, since much evidence exists associating past climate changes with solar activity and changes in the Sun's position in the Galaxy. So accepting Svensmark's findings for the moment, let's return to the passages of the Sun through the spiral arms of the Milky Way. If the Sun plied through at least four spiral arms in the last 500 million years, is there climatological evidence that corresponds to such events? A growing consensus of researchers say that there is. The geologic record is faithful at registering extended periods of cold (called "icehouses") and hot temperatures ("greenhouses") lasting for tens of millions of years. Isotopic evidence in ice cores and biochemical sediments all point to at least four periods of extended cold in the past 500 million years.[32]

Of course, the placement of the spiral arms with respect to the Earth's orbit (in addition to how long-lived the spiral pattern is, which is perhaps several billion years) will introduce some uncertainties into the calculated arm-crossing times. For example, if we assume each spiral arm is roughly 2,500 light-years wide, the time it takes for Earth to complete one crossing depends on the Sun's "angle of attack" to the curvature of the spiral arm. For example, passage through the Perseus Arm was relatively brief, on the order of 40 million years, because the Sun's trajectory with respect to the arm was not as longitudinal as it was with the Norma-Cygnus and Scutum-Crux arms, which each took about 80 million years for the Sun to traverse. There are other possible influencing factors that could come into play as well, such as encounters with Giant Molecular Clouds or gravitational disturbances that could alter the Sun's galactic orbit, impact the Earth's climate, or trigger other unpleasant consequences. However, let us stick to spiral arm crossings for now.

Scientists love periodicities that correspond with other periodicities, and the correlation between ice age epochs and the passage of the Sun through the Galaxy is certainly enthralling. How much more intriguing would the idea be, then, if these two cycles were also reflected in mass extinctions? Unfortunately, the evidence for this provocative notion is not very good. For there to be an accordance between ice age epochs and spiral arm crossings, the spiral pattern speed (or how fast the spiral pattern itself rotates about the galactic center) must be between 14 and 17 kilometers (8.6 and 10.5 miles) per second. For mass extinctions and spiral arm crossings to correlate, the spiral pattern speed would have to be closer to 19 kilometers (12 miles) per second.[33] But at that velocity, the association with arm crossings and ice ages becomes less tenable. Take another look at figure 12, which shows the spiral arm crossings of the Sun. The spiral pattern speed used to calculate the crossings is 14.4 kilometers (9 miles) per second. A distribution of five mass extinctions, as noted by the X marks, shows only one obvious hit, in the Perseus Arm. The others, with a little arm waving, might be said to be "close." Given that proviso, we might cautiously agree that mass extinctions *may* be periodic and *may* have something to do with the Earth passing though the spiral arms, but obviously, more research needs to be done to resolve the matter one way or the other.

The history of extinction is in itself a fascinating study dating back at least to the late seventeenth century, when English physicist Robert Hooke made microscopic studies of marine fossils and concluded that they were the remains of once-living organisms that had since evolved into new present-day species. Many of Hooke's contemporaries, however, who held the conventional religious belief that Earth was only 6,000 years old, found the idea repugnant. Why would God "re-create" an already perfect creation, they asked? In the words of English naturalist John Ray, it was a "dismembering of the universe" that made it imperfect.[34] But the idea would not go away, and by 1796 the French anatomist Georges Cuvier had classified numerous fossils that did not resemble the ossified forms of living animals. Nevertheless, many scientists continued to resist. Not until the Earth was explored to an extent that ensured that the absent animals weren't concealing themselves or had somehow been geographically isolated (à la Jules Verne's *Mysterious Island*) did the notion of extinction gain firm ground.

Mass extinctions—the sharp, sudden decline of most taxonomic groups living in a particular geologic period—might still be considered a contemporary adjunct to the extinction literature of the eighteenth and nineteenth centuries, even though the idea was initially given a cool reception. Its scientific fathers are many, but one of the most noted is Scottish geologist Sir Roderick Murchison. In the early 1840s, Murchison conducted a survey of the fossils and rock layers of the Ural Mountains and discovered a geologic time period sandwiched between the Carboniferous era (between 360 million and 290 million years ago) and the Triassic period (250 million to 208 million years ago). He named the new period the Permian, after an ancient kingdom called Permia, which once thrived west of the Urals. What made the Permian assemblage of fossils distinctly odd was that, while the lower, older rock sediments contained an abundance of fossilized life-forms, they were entirely absent in the higher, younger Triassic sediments. It was as if thousands of species had been wiped off the face of the Earth. Today, this dearth in the fossil record has been linked to what is known as the Permian-Triassic extinction event, or the "Great Dying," as it is often called. It stands as Earth's most extreme mass-extinction episode, with nearly 96 percent of all marine life-forms and 70 percent of all terrestrial life-forms obliterated.

In all, there have been at least five major extinction events and several minor ones. The cause of most is thought to be due to the glaciations of continents, but volcanism and climate change have also likely contributed. Moreover, there may be a pattern to these extinctions and, hence, a new and unsettling chapter in the evolution of life on planet Earth.

Back in the mid to late-1980s, David M. Raup and the late John Sepkoski Jr., of the department of geophysical sciences at the University of Chicago, published a series of papers demonstrating that major extinctions of fossil families of marine vertebrates, invertebrates, and protozoa occurred at intervals of 26 million years. They had no idea what caused the extinctions, but they were fairly certain that the probable cause was extraterrestrial in nature because, as they stated in their findings, "purely biological or earthbound physical cycles seem incredible, where the cycles are of fixed length and measured on a timescale of tens of millions of years."[35] One possibility, they speculated, was that the passage of the solar system through the Galaxy's spiral arms dramatically increased the comet flux near the Earth (an idea originally proposed in 1983 by planetary astronomer Eugene Shoemaker). At the time, ice age epochs spawned by spiral arm crossings had yet to be considered.

Based on this evidence, particle-physicist-cum-astrophysicist-cum-geophysicist Richard A. Muller at the University of California, Berkeley, theorized in 1985 that the periodicity might be caused not by spiral arm crossings but by another gravitational disturbance, one with a more ready-made cycle: the passage of a hypothetical low-mass star through the vast Oort Cloud of comets in the outer solar system. The idea was that this insidious star, which Muller called Nemesis, would sweep up untold numbers of comets in its wake and hurl them toward the inner solar system every 26 million years or so, the bombardment of which would result in periodic extinctions on Earth.[36] The theory seems plausible enough until you consider that, to date, infrared surveys for such death stars in the immediate vicinity of the Sun have turned up nothing. There are a number of new nearby stars on the solar neighborhood roster now—all more distant than the Sun's nearest stellar neighbor, Proxima Centauri—but none that could be remotely considered the Sun's notorious companion.

Still, cycles of mass extinction might yet be found in the geologic record if you know where to look. In 2005, a new analysis of the data by Muller and a graduate student revealed something intriguing—a distinct 62-million-year signal.[37] Again, no explanation for this periodic mass death knell was forthcoming. Perhaps, again, it was the intermittent passage of the solar system through interstellar molecular clouds affecting climate by varying cosmic-ray flux, or stimulating oscillation modes in which volcanic plumes reached Earth's surface at that period, or from an unknown "Planet X" perturbing comets. Whatever it was, however, a companion star wouldn't work in this case, because a stellar orbit with a 62-million-year period around a host star with a mass as large as the Sun's would be too unstable and susceptible to being diverted by passing stars.

Then, in August 2007, a new explanation for the 62-million-year cycle cropped up in a paper by astronomers Mikhail V. Medvedev and Adrian L. Melott of the University of Kansas. In short, they asserted that the motion of the Sun in the Milky Way, as well as the motion of the Milky Way itself through the intergalactic medium, could have serious repercussions for biodiversity on Earth.[38]

Astronomers have long known that the Milky Way is plowing through the intergalactic medium at a speed of about 200 kilometers (124 miles) per second in a direction north of the galactic plane toward the Virgo Cluster of galaxies. Just as the solar system is bounded fore and aft by shock fronts where the solar wind encounters the interstellar medium, the supersonic speed of the Galaxy's motion through the intergalactic medium creates a bow shock in the direction in which it is moving and a trailing termination shock in the opposite direction that displaces the center of the Milky Way's surrounding halo at least 200,000 light-years south of the Galaxy's midplane.

How could this possibly be relevant to mass extinctions? As noted earlier, as the Sun moves around the center of the Galaxy, it also oscillates in vertical distance north and south of the galactic midplane. The total amplitude of the oscillation is about 230 light-years and its period is approximately 64 million years. As the Sun approaches the peak of a northward oscillation, it lies closer to the Galaxy's leading bow shock, and the solar system is more exposed to the intergalactic medium. Protons and other nuclei can be accelerated by the bow shock, becoming high-energy cosmic rays capable of penetrating the galactic winds generated in the halo. According to Medvedev and Melott, the Sun's northward excursion causes a fivefold increase in the extragalactic cosmic-ray flux, and *this*, they claim, could adversely affect Earth's biodiversity on timescales of 60 million years.

This mechanism had never been considered as a candidate for producing mass extinctions because it was thought that the effect on the cosmic-ray flux was too insignificant. Like the Earth, the Galaxy's disk is shielded from extragalactic cosmic rays by its own magnetic field. But with the placement of the solar system nearer the bow shock north of the galactic plane, facing into a steady stream of accelerated cosmic rays as it were, the number of extragalactic cosmic rays reaching the Earth's upper atmosphere is not trivial.

Medvedev and Melott propose a grim menu of direct and

indirect mechanisms by which extragalactic cosmic rays might affect Earth's biodiversity, though they do not advocate one over another. These include: increased radiation damage, cancer, and genetic mutations, even among deep-sea and deep-Earth organisms; significant climate change induced by prolonged cloud cover from cosmic-ray ions; increased lightning discharges at lower altitudes, also brought on by cosmic-ray ionization, which in turn produces oxides of nitrogen that can precipitate into acid rain; and mutagenetic effects of solar ultraviolet radiation brought on by the aforementioned oxides of nitrogen compromising the ozone layer.

In a news article that appeared in the October 2007 issue of *Physics Today*, Melott makes it clear that he and Medvedev do not claim that increased extragalactic cosmic-ray flux is solely responsible for declines in biodiversity or major extinctions. "What we do suggest is that the increased cosmic-ray flux induces a periodic stress—each time lasting about 10 million years—that increases the vulnerability of the biosphere to whatever else might come along."[39]

In terms of natural events, "whatever else that might come along" could include—in addition to getting zapped by high-energy extragalactic cosmic rays—anything from an unexpected encounter between the Earth and a Giant Molecular Cloud, the onset of an ice age, or a major asteroid impact. After all, it stands to reason that if life, bacterial or something more advanced, is already suffering from a planetwide blight or adverse climate change, a comet impact probably wouldn't help matters. Perhaps the greatest mass extinctions of the past, like the Permo-Triassic event, were attributable not to one cause but several. The same might be said about future mass extinctions. Imagine our civilization already brought to its knees by some humanmade nuclear or biological holocaust, or entrenched global warming, and then along comes an asteroid.

Evolution, even for those who staunchly object to it, is not a concept we usually associate with extinctions of any kind, since the verb form implies development, diversification, and dissemination, not the denouement of life. But extinctions are, nonetheless, necessary components of evolution. It is therefore instructive to read Charles Darwin's views about the sudden extermination of whole groups of species, to which we may now add ourselves. In *On the Origin of Species*, Darwin argues that such extinction events accord well with his theory of natural selection, since many unacknowledged checks and balances determine the ultimate fate of any species. He writes: "We need not marvel at extinction; if we must marvel, let it be at our presumption in imagining for a moment that we understand the many complex contingencies, on which the existence of each species depends."[40]

But what would Darwin say about the evolutionary consequences of an ice age epoch being brought on by the Sun's roller coaster ride through the Galaxy or by the Galaxy's motion through extragalactic space? How would his survival-of-the-fittest argument accord with the planet-killing impact of a comet or an asteroid? And how would he view natural selection in light of the fact that many of the progenitors of the foraminifera, mollusks, lemurs, pachyderms, and such—that he so assiduously studied during his world voyage aboard the *Beagle*—probably owed part of their physical traits to the mutagenetic effects of supernovae or gamma-ray bursts? These are events that themselves may have wiped out life on some other planet. If you could have one of those fantasy conversations with anyone you choose, living or dead, I would imagine Darwin, upon hearing such questions, would scowl a little and reply in a huff: "Pooh!

My theory still stands despite the influence of astronomical factors. Extinctions and mutagenetics are simply cogs, chance cogs though they may be, in the workings of evolution. Didn't you read *any* of my books?"

If he added anything to this—and well he might, given what is known today about Earth's rich astronomical history—I suspect he would nonetheless keep within the lines of what he said in the previous paragraph, with, perhaps, two additions, which I place in italics: Evolution, with all its countless behind-the-scenes machinations, *operating not just on the scale of the stars but on the scale of the universe*, remains the most natural means of sorting out the many processes of chance and necessity affecting the outcomes of life—*and extinctions*—anywhere and everywhere.

CHAPTER 7

KEEP WATCHING
THE SKIES!

SETI (the Search for Extraterrestrial Intelligence) is now a fully accepted department of astronomy. The fact that it is still a science without a subject should be neither surprising nor disappointing. It is only within half a human lifetime that we have possessed the technology to listen to the stars.
—Arthur C. Clarke, prologue to *Childhood's End*

In the original 1898 version of *The War of the Worlds*, British author H. G. Wells describes the first of the invading "cylinders" from Mars as falling "somewhere on the common between Horsell, Ottershaw, and Woking," a location that is about 35 kilometers (22 miles) southwest of the heart of London. The site, described by Wells as being not far from the sandpits on Horsell Common, had gouged out a deep hole and flung out sand and gravel in every direction. Today, you can actually see a drowned sandpit at this location on Horsell Common. Though fictional, the setting has come to be regarded, even revered, as the place where the Martians landed. Over the years, the crater would be relocated around the world in various retellings of the story, but no matter where the first cylinder landed—it could be anywhere, really—it always served as the same metaphor: the catalyzing moment when the human race learned it wasn't alone in the universe.

I had read the novel in my youth (originally, truth be told, in the comic book series *Classics Illustrated*) and watched the 1953 movie (with Gene Barry playing the starring role of the hapless astronomer Clayton Forrester) so many times I practically knew all the dialogue by heart. But a few decades down the road, I had shelved the details of the original plot until one autumn in 2000, when I accompanied my wife to Woking to visit her sister Pamela and twin nephews. The afternoon we arrived, still much jet-lagged, Pamela suggested a calming stroll on Horsell Common, a large, wild park area that was a favorite running-about place of the boys, who were a very lively three years old at the time. It turns out that Wells had, in fact, lived and worked for a time near Woking on Maybury Road, and it was there that he had written *The Invisible Man* and *The War of the Worlds*. According to local historians, Wells learned to ride a bicycle in Woking and afterward rode around the area noting places and people that were to be destroyed by the Martians.[1] No wonder the first wave of the marauding Martians landed there. It was his own backyard. In a flash my jet lag was forgotten. "Let's go!" I said.

After parking the car, we made our way along a narrow, well-trodden path through the dense, wild woods. It was a cold, clear afternoon. The Sun cast a rich amber light, and the shadows were already long. We soon emerged into a brilliant concave clearing about the size of a baseball infield: this was the famous sandpit.

I don't know what I was expecting, but obviously I was expecting more than what I saw before me. The pit was crowned on one side by a broad "beach" of tawny sand and gravel that sloped gently down to an oxygen-starved pond of standing turbid water on which floated sticks and scum. In some places narrow spits of spongy ground extended along the pond's edge, from which sprouted some low, scraggly brush and gaunt grass. The notorious landing site of Wells's Martian vanguard force was neither deep nor sinister. It was a mudhole.*

Apparently, the sandpit was quite a bit deeper back in Wells's time. In fact, sand and gravel had been extracted from the Common's pits for centuries.[2] Today, however, it appeared to be nothing more than a low sump in the middle of a semi-wild English wood—albeit a lovely wood, particularly on a crisp, clear, sunny afternoon. "It looks better," my sister-in-law remarked, "when it has more water in it." I snapped a few pictures, trying to imagine this unassuming spot as Wells envisioned it, the place where locals gathered to watch, entranced, as the top of the mysterious cylinder from Mars unscrewed itself and fell heavily to the ground with a metallic ring, and the hulking bodies of the Martians, with slathering, lipless mouths, heaved themselves through the black opening into the slanting afternoon light. Perhaps the setting was better at sunset, I thought, just as it was in Wells's book. The horror would come in the deepening twilight when the milling crowds, at first silhouetted against the dusky sky, would suddenly cry out and scatter like panicked rabbits as the terrible heat ray mowed them down as they fled. What, I wondered, would be the last thoughts of a people unexpectedly exterminated by a race of beings from another world?

It was just such a seed that was planted in the public's imagination back then by Wells's novel, and science fiction and science itself have nurtured the rapid outgrowth of that seed ever since. *The War of the Worlds* was the first work of its kind to make us look

---

*To be fair, Horsell Common, though not the sandpit itself, features some important Bronze Age burial sites dating back to 1500 BCE.

up and wonder if armies, not from other countries but from other planets, could invade and overpower our entire world, all within days, before any country could rise above its own prejudices and work with others to combat a new and dispassionate enemy from the stars. The theme touched upon a nameless, deep-seated fear. Unfortunately, this abstract became all too representational during the world wars of the twentieth century, when attacks from an invading country created apocalyptic scenes in places, and in psyches, that many people considered inviolate.

What made *The War of the Worlds* truly frightening then and today was the idea that intelligent life (intentions unknown) might *come to us* and land anywhere at anytime on Earth. Why not Horsell Common? It's not any more unlikely a place than Grover's Mill, New Jersey, where actor Orson Welles and fellow scriptwriter Howard Koch staged the 1938 radio version of the Martian invasion that terrified many listeners who thought they were listening to news bulletins reporting the landing of multiple cylinders from the Red Planet.

The gripping question is, could such an invasion actually happen? Of course, most sober-minded folk dismiss the idea, and for good reasons. But it's worth keeping in the back of your head that the Earth was once vast enough to insulate numerous tribal societies from modern civilization, at least up until the beginning of the twentieth century with the ramping up of territorial expansionism and manifest destiny. Even today, new tribes are allegedly being discovered in countries such as Peru and Brazil.[3] If expansionism is part of an alien society's culture, as it is with our own, perhaps the idea of being visited by a race of extraterrestrials is not so far-fetched, and their presence not so far off the radar of reality, especially if said society is a few thousand years ahead of us, technologically speaking.

However you consider it, from the viewpoint of science or science fiction, philosophy or theology, the question of whether or not intelligent life exists on worlds other than our own is a fascinating intellectual exercise that continuously loops back on itself like a Möbius strip, as the same ground gets repeatedly covered without ever arriving at a satisfactory conclusion. The question itself entails numerous other questions, including the evolution of life in selective, and different, environments other than our own, and the metaphysical variables inherent in what is termed life and consciousness. As Carl Sagan put it, looking far down the road as he was wont to do: "We have not witnessed the evolution of biospheres on a wide range of planets. We have not observed many cases of what is possible and what is not. Until we have had such an experience—or detected extraterrestrial intelligence—we will of course be enveloped in uncertainty."[4]

The question of life on other worlds is a very old one. Just how old, no one can say for certain, but most scholars agree that it dates back to the very roots of human sentience, when self-made, anonymous philosophers roamed the Earth before the advent of writing. In terms of actually going down on record, one of the first was the Greek philosopher Epicurus in the fourth century BCE: "There are an infinite number of worlds both like and unlike this world of ours.... We must believe that in all worlds there are living creatures and plants and other things we see in this world."[5] The Roman poet and philosopher Lucretius, in his poem "The Nature of Things," echoed Epicurus: "[O]ther worlds must exist in other regions, and different races, both of men and beasts."[6]

Beginning around the mid-1500s and extending well into the seventeenth century, European Renaissance scientists made formidable strides toward unifying astronomy, chemistry, and biology under one roof, while at the same time raising the imposing bar of cosmic knowledge by promulgating and expanding upon the Copernican notion that Earth was not at the center of the universe; in fact, it wasn't even close. The telescopic observations by the first scientists such as Galileo, and subsequently the Dutch physicist Christiaan Huygens, revealed that planets were worlds unto themselves, complete with their own atmospheres and, in some cases, moons (or "satellites," as Johannes Kepler referred to them). Given that such observations were often deemed blasphemous by the Catholic Church, it's not surprising that along the way there were also a few casualties, most notably the Dominican friar, philosopher, and mystic Giordano Bruno. In 1593, sixteen years before Galileo began making observations with his telescope, Bruno was imprisoned by the Roman inquisition and eventually burned at the stake for, among other heresies, writing that other worlds were "if not more nobly, at least no less inhabited and no less nobly" than our own.[7] Of course, in retrospect, there was no need to burn alive someone for his ideals, as those ideals were already proliferating among the lower classes anyway. Undermined by other so-called great thinkers and theologians of the day, theirs was a reflexive and putative reaction that resonates to this day.

The list of scientists, philosophers, and scholars who agreed with Bruno down through the ages is a long one: Kepler, Huygens, Richard Bentley, Immanuel Kant, William Herschel, and Benjamin Franklin are just a few. As for those who have since speculated about, or advocated for, intelligent life on other worlds, the number would be the total of practically everyone who has lived in the last 200 years. The question, which has gone from philosophical musings to scientific scrutiny, has the greatest ramifications for humankind today.

No one appreciated the idea of life on other worlds more than

Wells, who used the question as an effective vehicle for *The War of the Worlds*. He was wise to choose Mars as Earth's alien antagonists because, at the time, astronomers were just beginning to earnestly debate whether or not the Red Planet harbored life—plant, animal, or otherwise. The planet's two tiny moons, Deimos and Phobos, had been discovered only in 1877, twenty-one years before the novel was published, as Earth and Mars approached the points in their respective orbits when they were nearest each other, a configuration known as opposition.* The minimum distance that year—a mere 56 million kilometers (35 million miles)—was about as close as Mars ever gets to Earth, and the rare opportunity encouraged astronomers around the world to turn their telescopes, big and small, toward Mars. In the years leading up to *The War of the Worlds*, Mars reached opposition on at least two occasions, in 1892 and again in 1894. Both were almost as favorable as that of the 1877 opposition (particularly the one in 1892). An opposition in 1896, though far from ideal (the planet's minimum distance from Earth was 84 million kilometers, or 52 million miles), nevertheless garnered much attention from astronomers and, no doubt, loomed large in the public imagination when *The War of the Worlds* first appeared as a series of stories in *Pearson's* magazine in 1897. The timing for such a work couldn't have been better.

Whole books and countless articles were written about Mars during this period, and, as a result, the planet has taken on an almost mythical quality. Mars, which had been noted in the most ancient of astronomical records primarily for its striking reddish hue, was being discovered anew and endowed with all-too-

---

*During opposition, Mars, or any planet outside the orbit of Earth, rises in the east as the Sun sets in the west and is said to be "opposite" the Sun in the sky. At that time, because it is nearest Earth, the planet's disk is larger and surface details, which are otherwise too faint to be seen at other configurations, become more visible.

terrestrial characteristics. Martian surface features were christened with quaint earthlike complements such as Niester Isthmus, Huggins Bay, Terby Sea, Noble Cape, Lake Schiaparelli, and the Mitchell Mountains.[8] And because the optical resolution of telescopes improved dramatically during this period, new and compelling features—interpreted as clouds, mists, spots, lineaments, and such—began appearing in sketches and observing reports. In sum, the observational chronicles of Mars during the late nineteenth and early twentieth centuries brimmed with insights and discoveries, as well as wild speculations. It must have been a heady time to be an observational astronomer; even a topical search of the archives today turns up report after report about the planet's continent-like features, how they appear to be dark or dark green during Martian spring but fade to a neutral tint (some describe it as brown) in Martian autumn. These dark areas were inferred by some scientists to be lichen-like vegetation that could survive the Martian winters. Observers also noted the waxing and waning of the north and south polar caps as the planet wheeled through Martian summer and winter and freely speculated about the presence of water, lakes, and even oceans. Mars became a world of possibilities that essentially galvanized the age-old discussion of extraterrestrial life for decades to come.

By far the most intriguing and sought-after features on the planet's surface, however, were the apparent presence of straight dark lines first mapped between 1877 and 1881 by Italian astronomer Giovanni Schiaparelli. Earlier Mars geographers, or "aerographers" as they were often referred to,* had plainly noted dusky shape-shifting surface features, but none specifically mentioned or sketched linear markings as such.[9] As is widely known today, Schiaparelli referred to them in his observing reports as

---

*After Ares, the Greek god of war (Mars being the Roman version).

canali, which is Italian for "grooves" or "channels."* Unfortunately, rash English tongues created confusion by mistranslating *canali* into "canals," which plainly implies an artificial waterway. With that academic miscue, Pandora's box flew open, and the debate was on as to whether *canali* were real or illusionary, natural or artificial, or attributable to life, either rudimentary or of some higher order. Moreover, the discussion wasn't restricted to the province of the scientist. Any amateur astronomer equipped with a fairly decent telescope could look for these same features, real or imagined, from his or her own backyard.

Judging from articles that can be accessed at online archives such as NASA's Astrophysics Data System, a great many astronomers of the day came down in favor of the canals being some perception effect similar to Mach bands, in which two adjoining dark and light regions create the illusion of a distinct grayish band (or line) between them. As Irish telescope mirror maker and observer C. E. Burton put it, the delicate markings are "boundaries of differently tinted districts than independent streaks."[10]

Edward Walker Maunder, namesake of the famous Maunder minimum of sunspots noted earlier, brought up a geometric objection in 1888: "It need hardly be pointed out that actual straight lines on the surface of the planet could only appear to be straight when seen near the centre of the disk. Everywhere else they would be seen as curved; near the limb they would appear strongly curved. It is therefore impossible to accept these canals *as delineated* [Maunder's italics], as representing real markings on the surface of the planet."[11]

Then there were other astronomers who were just as adamant that the canals were artificial. Leo Brenner, who, truth be told, was one of the more controversial astronomers in his day, wrote that

---

*See G. V. Schiaparelli, "La Vita Sul Pianeta Marte," *della rivista* "Natura ed Arte," March 1895.

the canals could not have been made by nature: "Everything leads to the conviction that this network [of canals] is artificial."[12] Brenner was not alone in his convictions. French astronomer Nicolas Flammarion wholeheartedly supported the idea that Mars was inhabited by intelligent life, most likely more advanced than that of Earth. American astronomer William H. Pickering, too, suspected that at least vegetative life existed alongside the canals (although Pickering also thought the Moon held some form of life as well).[13] Pickering even pointed out that the alleged Martians could be signaled with a mirror one-half square mile in area. Such a signal, he wrote, "would, therefore, be dazzlingly conspicuous to Martian observers, if they were intellectually and physically our equals."[14] And English astronomer and science popularizer Richard Anthony Proctor, in his book *The Universe of Suns*, began a chapter called "Life in Mars" this way: "All that we have learned about Mars leads to the conclusion that it is well fitted to be the abode of life."[15]

But such speculations were just a prelude to American astronomer Percival Lowell's contributions to the Mars mythology. Lowell, who was born into an aristocratic Boston family, had the money and means to establish his own observatory in the mountains of Arizona in 1894, from which he scrutinized Mars as a microscopist might scrutinize an amoeba. During the opposition of 1894 and 1895, Lowell, along with Pickering and another American astronomer, Andrew Ellicott Douglass (who was later to become a respected pioneer in dendrochronology), reported observing 183 canals.[16] For the next fifteen years, Lowell took thousands of photographs of the planet, some of which he claimed showed the elusive canals, though they never appeared as clearly as they did in his sketches. In all he plotted over 500 of them, mapped numerous "oases" where they converged, and reported how they changed with the Martian seasons.[17] His observations took the idea of artificial canals and life on Mars to a level that fired up the imaginations of the general public not only of that time, but for years to come.

At the height of all the Martian canal kerfuffle, the specter of a Martian invasion was inadvertently set loose upon the northeastern coast of the United States when Orson Welles staged his radio version of *The War of the Worlds* on the evening of October 30, 1938. In another interesting convergence of celestial and terrestrial events, Mars had just come to opposition the year before, when it was a very bright object in the evening eastern sky and much sought after by amateur astronomers and the general public alike. Astronomer Robert Barker, a fellow of the Royal Astronomical Society, reported in *Popular Astronomy* about regions of varying tones that "hint at profuse vegetation" as well as many canals that are "clear cut and definite." He also included a detailed map showing these same features.[18]

The Red Planet, then, was a high-profile item in astronomy news in the autumn of 1938, and whether for that reason, or because *The War of the Worlds* was celebrating its fortieth anniversary, or just because Mars by then had such a notorious reputation, Welles and Koch decided that their adaptation would have a more visceral impact if it were removed from its original Victorian backdrop and set in present-day America. They jacked things up even further by choosing to produce the drama in the form of a series of news flashes that would periodically break in on a program of contemporary dance music, with each bulletin increasing in emotional intensity. At some point, the news flashes became ongoing, with live reports from locations where the Martians had landed and were apparently blasting away at hapless country folk who had come to have a look-see, as it were.

What happened next has been well chronicled in numerous books, articles, and Web pages, so a summary here will suffice.[19] Telephone trunk lines into the Trenton, New Jersey, police headquarters were jammed by panic-stricken citizens who demanded to know what was going on. Newspaper reporters from Detroit and Washington, DC, called, asking for details of the invasion. Roads in the vicinity of the "invasion" site, Grover's Mill, were jammed with people trying either to escape with their lives or to reach the scene of the invasion to satisfy their own curiosity. Callers from as far away as Boston reported seeing the flames of ravaged New Jersey, and telephone lines from California were tied up with people worried about the fate of their friends and relatives back east. Some people actually reported seeing the Martian tripods themselves striding across the land. In New York City, people ran into Central Park spreading the news of the invasion, and some families gathered there or in the streets to await their doom.

Fortunately, the hysteria died away almost as quickly as it arose, but the fear that the broadcast sparked was never to be forgotten and lives on today in the annals of radio history, and in more than a few sociological studies. Repeat broadcasts in Santiago, Chile, in 1944 and Quito, Ecuador, in 1949 prompted serious incidences of widespread panic, riots, and even a few deaths. Princeton University psychologist Hadley Cantril, who in the 1940s published a well-known study of the public reaction to the radio broadcast, attributed the panic to, among other things, the unusual realism of the program and the public's acceptance of radio as a legitimate medium for important announcements.[20]

Speculation about the reality of Martian canals extended up to 1965; however, save for a few Lowell Observatory holdouts, most astronomers by then had come around to the idea that Mars was a lifeless, frozen desert world.[21] This seemed to be confirmed when *Mariner IV* examined the planet's surface from a distance of 14,000 kilometers (8,700 miles) in July 1965. No canals were evident, but numerous impact craters, like those found on the Moon, were. The planet's atmosphere, astronomers concluded, was much thinner than had been anticipated; otherwise more meteors would have burned up before reaching the surface. Even then, there were holdouts for at least elemental life on Mars, if not something more advanced but as yet undetected. Carl Sagan pointed out that had *Mariner IV*'s television camera been pointed toward Earth from that same altitude, its limited resolution would have revealed no sign of life, not even a city the size of New York.[22]

The final blow for the prospect of intelligent life on Mars was delivered by the *Mariner IX* spacecraft beginning in November 1971. With resolutions between 100 and 1,000 meters (330 and 3,300 feet), the photographs revealed a surface rich in geologic detail—heavily cratered terrain, large circular basins, vast dune fields, complex canyon systems, and expansive volcanoes. But, again, no canals. In fact, the improved photographic resolution clearly pointed to a world that was in an intermediate evolutionary stage between the Earth and the Moon.[23] As fascinating as the results were, there was no denying that one of the most enduring astronomical questions concerning life on Mars, or at least intelligent life, had been definitively answered. We need never worry again about invaders from Mars.

Still, even though Mars has been checked off the list, hope for extraterrestrial intelligence springs eternal. After all, hordes of nearby stars remain that may shelter possible habitable planets. "Our Sun is only one of several thousand million suns in our universe," wrote Robert G. Aitken, associate director of Lick Observatory in California, in 1925. "Must there not among them be many having planets that are not only capable of sustaining life, but that have intelligent beings dwelling upon them?"[24] It's the same old odds-on assumption astronomers use today to convince themselves that there must be an abundance of life-bearing planets out there, and that circling every sunlike star should be at least one or two blue planets with abundant surface water.[25]

This was also the assumption that inspired the first modern step forward in the search for extraterrestrial intelligence just as the Sun was setting on the idea of Martian life-forms. In April 1960, astronomer Frank Drake turned the 85-foot Howard Tatel radio telescope at the National Radio Astronomy Observatory in Green Bank, West Virginia, toward two nearby sunlike stars, Tau Ceti (12 light-years away) and Epsilon Eridani (10 light-years). For six hours each day, in what he called Project Ozma, he tuned into first one then the other, listening in at a frequency of 1420 MHz, or at a wavelength of 21 centimeters, which corresponds to the wavelength of neutral hydrogen, the most abundant element in the universe. Seven months earlier, physicists Giuseppe Cocconi and Philip Morrison of Cornell University suggested in a paper in *Nature* that because of its universality, the frequency of neutral hydrogen would most likely be the optimal channel to be used by other advanced extraterrestrial civilizations to make contact; therefore, a search around that frequency band would be the best place to begin a search for ET. They also predicted that the signal would most likely be in the form of modulated pulses on the order of seconds.[26]

The search began on an auspicious note when, on the very first day, not too long after they had pointed the telescope toward Epsilon Eridani, Drake and the telescope operator were startled to hear a series of rapid staccato noise bursts that sounded like "ch-ch-ch..." coming from the loudspeaker that was tied into the telescope's receiver output. The bursts occurred at a precise rate of eight times per second—very similar to what Cocconi and Morrison had predicted. To determine if the sound was really coming from the star, they moved the telescope off Epsilon Eridani, whereupon the sounds went away. Excitedly, they slewed back on to the star again but were met with disappointment. The sounds should have returned if they were truly coming from the star, but they did not. In fact, a week passed without the signal's recurrence. All this time, Drake couldn't tell whether the initial bursts were brief signals from Epsilon Eridani or if they originated on the Earth and just happened to cease when the telescope was pointed away from the star. Unfortunately, it turned out to be the latter. Ten days later, Drake and the telescope operator were abruptly blasted again by the noise bursts. This time, however, the sounds had also been detected just as strongly by a second smaller receiver connected to a simple horn antenna, which had been employed to pick up Earth-bound interference. The "signals," then, were not from Epsilon Eridani but, perhaps, some high-flying military aircraft. Project Ozma continued monitoring the two stars until July 1960, but the rest of these signals, as they say, was noise.[27]

Since then, there never has been a signal that could be attributed to extraterrestrials, though there have certainly been a few interesting cases. The most famous of these is today referred to as the "Wow!" signal, which was picked up by a radio telescope called the Big Ear on August 15, 1977, operating at Ohio State University. The signal was noted by Jerry Ehman, a professor at Franklin University, whose job it was to review the Big Ear's observations. Signals were recorded on a continuous printout sheet using a long list of numbers and letters. The numbers (1–9) represented the signal level above the background noise, and the letters represented increasingly stronger signals (with A being the weakest and Z the strongest).

The signal string recorded that evening was "6EQUJ5." This

Figure 13. The Howard E. Tatel Radio Telescope, built in 1958 for the National Radio Astronomy Observatory in Green Bank, West Virginia, was used by Frank Drake in 1960 to make the first scientific attempt to detect signals from an extraterrestrial civilization. Courtesy of NRAO/AUI.

indicated that the signal was well above the background noise of radio interference (6) and grew in strength from level "E" to a peak of "U" before dropping down to "J," but was still above a background level of 5. Moreover, the signal was not intermittent but continuous and came and went within a time span of 37 seconds, which was the "scan time" of the Big Ear beam for any given point in the sky. Hence, any signal from ET would smoothly rise, peak, then decrease over a period of 37 seconds. Ehman, who spied the signal string a few days later during his review of the printouts, circled it and wrote "Wow!" in the margins.

In years following, several attempts were made to redetect and isolate the "Wow!" signal, but to no avail.[28] The signal's origin on the sky, in the southern reaches of the constellation Sagittarius, has never been identified with any known source, and today, whatever caused it is still somewhat of a mystery. Astronomers are pretty certain, however, that whatever the source, it was probably not from extraterrestrials.

By this time, Frank Drake had made a name for himself with another, more important, contribution to SETI: a means of estimating the number of planets in the Galaxy with intelligent technological civilizations. His equation, which still bears his name, is, aside from $E=mc^2$, probably one of the more popular mathematical statements in science, but it is also one that is open to the widest interpretation.

Readers already familiar with the Drake equation will recognize the following mathematical hieroglyphics:

$$N_{ac} = R^* \, f_p \, n_e \, f_l \, f_i \, f_c \, L$$

Each term represents a probability in the evolutionary step of a galaxy, its stars, and possible life and possible intelligent life. Hence the number of alien civilizations (N_{ac}) communicating in the Galaxy depends on (R^*) the average star-formation rate of suitable stars, (f_p) the fraction of those stars with planets, (n_e) the number of those planets that are earthlike, (f_l) the fraction of those planets that go on to develop life, (f_i) the fraction that go on to develop intelligent life, (f_c) the fraction that develop a technology that can communicate with other intelligent life-forms in the Galaxy, and (L) the average lifetime of a technological civilization (i.e., one that doesn't exterminate itself). Multiply these numbers with any variables you like and you will get an answer between zero and millions or more. Drake's own value in 1961, was, by *Star Trek* standards, a very conservative 10 civilizations. He now believes that number is too small by at least a factor of 1,000 and could be a great deal larger still.[29]

A more current approach seeks not to estimate the number of worlds in the Milky Way that might harbor intelligent life but rather the probability of detecting radio signals from alien civilizations, a subtle but nonetheless different methodology.[30] Marko Horvat of the University of Zagreb, Croatia, expands upon the Drake equation by factoring in the logical constraints of interstellar radio communication. He sets out three propositions that would affect the probability of detecting radio signals from alien civilizations. First, if a signal is ever detected, it will in all likelihood *not* be an intentional signal—that is, one beamed specifically for the purposes of making extraterrestrial contact—but a random alien communication. Second, in order to pick up this unintentional signal we have to be listening for it during the interval when an alien civilization is employing radio communication, not before (when the technology hasn't been developed) and not after (when radio becomes for them obsolete, as it will for us eventually). Third, it is necessary to define a volume of space in which alien transmissions may be strong enough to be detected on Earth, assuming that our radio receivers are sensitive enough to detect alien signals.

Horvat considers a number of variables addressing these propositions, including the probability that our civilization is listening when another is transmitting, the volume of space around an alien transmitter in which its signal is detectable, the volume of space containing a certain number of alien civilizations, and the probable number of alien civilizations that will be within our communications range. To make his calculations more realistic, he includes two further restrictions that, in simple terms, might be stated this way: not all wavelengths can be monitored all of the time and not all areas of the sky can be monitored all the time. These are, of course, two restrictions astronomers must deal with when making any practical search for alien signals. In the first case, the ideal would be to listen to as many channels as possible, even though, in reality, only some can be tuned in. In the second case, it is not possible to scan all parts of the sky at once; more radio telescopes than exist today would be needed to do that. Horvat calls these restrictions the *wavelength ratio* and the *area ratio*, respectively.

Finally, he applies all the various values to derive 15 probability cases, some of which are more likely to be successful than others. The most optimal case (in which there is a 95 percent chance of at least one detection) is a volume around the Earth with a radius of 50,000 light-years occupied by 300 alien civilizations with a radio time of 2,500 years, and civilization life spans of 250,000 years. Reduce the number of civilizations to 100, and the probabilities of a successful detection fall to 63 percent; plug in Drake's original estimate of 10, and that value diminishes to 0.3 percent. Now consider the odds of successfully intercepting an intentional beam of radio emission that is very strong but lasts perhaps only minutes. Those odds would be negligible, to say the least.

Horvat, whose specialty is mobile computing and artificial intelligence, worked for six years on his paper, which deals strictly with the logical modeling and randomly determined mathematics of the problem. "I was careful to leave radio-astronomy questions to professional radio astronomers," he told me in an e-mail.

The length of time that a civilization remains radio "active" may be answered by our own civilization in a few centuries. Horvat agrees that even now fiber-optic cables, directed low-power microwave antennas, and radio and television programs being transmitted from satellites toward the Earth are changing our radio presence in interstellar space. "The usage of common radio will reach its peak then slowly drop," he says, "although when this exactly happens remains to be seen. I don't think we will abandon radio altogether...but as we use it less and less we will steadily become 'radio-quieter' and thus harder to detect by an alien SETI project."[31]

The same may be assumed about other hypothetical alien civilizations out there that eventually outgrow radio and turn to communicating by other means, perhaps using pulses of laser light that are many times brighter than the Sun and can be observed over interstellar distances. This is why, says Horvat, the search for extraterrestrial intelligence needs to cover as much sky as it can, and as fast and as sensitively as it can, using radio and optical telescopes. If the result of a survey is null, then its sensitivity needs to be increased, the range of detection extended, and the survey repeated. "The cycle will go on until we detect a signal," he says. "Naturally, that could be never."[32]

But not for want of trying. In 2005, after several years of planning, the nonprofit SETI Institute in Mountain View, California, in collaboration with the University of California at Berkeley's Radio Astronomy Laboratory, began constructing a new radio telescope complex called the Allen Array in Hat Creek Valley, some 300 miles northeast of San Francisco. As with other piggyback approaches to SETI, the Allen Array will conduct normal astronomical observations while SETI is along for the ride, except in this case, searches will be carried out 24 hours a day, 7 days a week, significantly speeding up the searching of targeted stars. When completed, hopefully within the next decade, the array will consist of 350 antennas, each 6.1 meters (20 feet) in diameter, enabling SETI researchers to eavesdrop on radio broadcasts from a million potential galactic civilizations as distant as 1,000 light-years.[33] The first 42 antennas of the array, which became operational in October 2007, were used to survey 20 square degrees around the galactic center. Such a swath of sky easily contains several billion stars, but these same stars are also 10,000 light-years or more away, meaning any signal would have to be very powerful to be detected.[34] As we shall see, though, we may be wise to be wary of ETs bearing potent radio signals.

One of the most practical traits of SETI programs of the past and today is their opportunistic nature. Two projects developed in the 1980s called SERENDIP (Search for Extraterrestrial Radio Emissions from Nearby Developed Intelligent Populations) utilized auxiliary receivers on radio telescopes that were performing observations for other projects. The SETI "agenda" merely piggybacked along in case ET might be transmitting interstellar greetings. SERENDIP is still operating today. In fact, it has an antenna mounted on the bridgelike receiver array structure that is high above the famous cereal bowl–shaped Arecibo radio dish in Puerto Rico. Essentially, the SERENDIP receiver scans whichever part of the sky the dish happens to be pointing toward. There is also a SERENDIP counterpart operating from the Parkes radio telescope in Australia, so the southern sky is covered as well. Together, Arecibo and Parkes cover the entire sky every three years looking for telltale recurring signals.

A similar commensal approach will soon be available using a future generation of radio telescope arrays that will be looking not purposely for intelligent signals from outer space but for the first galaxies and stars to form in the universe. And, more important for non-SETI astronomers, it will search for the oldest of all elements, neutral hydrogen, which filled the universe in the aftermath of the big bang. This is a matter of utmost importance to cosmologists trying to figure out why the early universe is so replete with structure in the form of weird-looking galaxies and massive black holes. In regard to SETI, as astronomers sweep the distant reaches of the universe for neutral hydrogen, they may inadvertently sweep up intelligent signals beaming from nearby inhabited stars.

Here's the cosmological agenda. What is currently accepted is that some 300,000 years after the big bang 13.7 billion years ago, the temperature permeating the hot plasma of the just-born universe had cooled to about 3,000°C (5,000°F), which is roughly half as hot as the surface of the Sun. This was cool enough, however, to allow protons to link up with electrons, which is what neutral hydrogen consists of: a single proton orbited by a single electron. Lowly as it is, it's still the most abundant element in the universe.

In a few more years, several new radio telescope projects—the Mileura Wide-Field Array (MWA), now under construction in Australia; the Low Frequency Array (LOFAR) being constructed across the Netherlands and Germany; and, further down the road in 2020, the Square Kilometer Array—will turn their capacious collecting areas toward the most distant reaches of the universe. This is where the earliest atoms of neutral hydrogen dwelled, humming away before they were ionized by light from the first stars.

Now, here's the SETI agenda. Because the expansion of the universe stretches all electromagnetic radiation to lower wavelengths (or frequencies), the faithful 21-centimeter line of neutral hydrogen won't be found at its usual spectral address of 1420 MHz but somewhere lower than that, between 80 and 300 MHz. This frequency range, then, is where these powerful radio telescope arrays will be listening. It just so happens, though, that this frequency range also coincides with the band over which our civilization is the most voluble and, perhaps, others too. The big idea is that while astronomers are listening in at these frequencies in their attempts to detect the telltale tick of primordial hydrogen, they'll also be attuned to picking up unintended radio "leakage" from potential earthlike civilizations out to tens of light-years and within a volume of space containing between 1,000 and 100 million stars.[35]

One of the astronomers proposing this radio eavesdropping approach to SETI is Abraham Loeb at Harvard University. Loeb says an optimal survey would search for a periodic signal from a selection of nearby stars. The modulation of any signals detected could be the result of the scanning pattern of an extraterrestrial military radar or the spin or orbit of the planet from where the signal is being emitted. This approach would still present limitations, however. "Observatories like MWA and LOFAR would not be able to detect a show like *I Love Lucy*," says Loeb, "but if any signal is detected, I'm sure that there will be plans for building more sensitive observatories." Loeb cautions, however, that the strongest signals emitted by our civilization come from military radars, not television broadcasts, and these are the ones that are most likely to be detected across interstellar space. Such a detection would have ominous implications. "The brightest ones might be the most militant," says Loeb. "My own preference would be that we do not reveal our identity because [they] might be hostile."[36]

Some readers with mental images of Steven Spielberg's endearing little lost alien in the movie *E.T.: The Extra-Terrestrial* may have a problem believing that beings more intelligent and more advanced than us could possibly be hostile. But where is it written that alien civilizations more technically advanced than Earth's are more benign or tolerant? After all, while we can claim to have more knowledge about the natural world and to be more

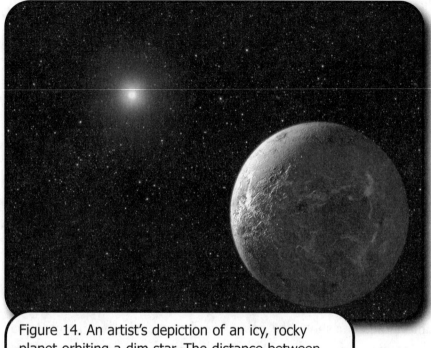

Figure 14. An artist's depiction of an icy, rocky planet orbiting a dim star. The distance between the planet and its host is about 450 million kilometers (280 million miles), three times the distance between Earth and the Sun. As such, this hypothetical planet is too cold to support life as we know it. Courtesy of NASA; ESA; G. Bacon (STScI).

technically advanced than, say, the ancient Romans, it is debatable whether we can also claim to be any more principled.

Some may ask, what is there to really worry about? Because of the great distances between the stars, it would probably take a century or tens of centuries for a hypothetical race of extraterrestrials to detect and then respond to our reply. That's indisputable, of course. But if they are already several centuries more advanced than us, then they have probably conquered space travel over great distances. (Imagine, optimistically, where Earth's technology of space flight will be in, say, 200 years.)* If proximity isn't too much of a problem, their "response" to our reply could conceivably be in the form of a visit or, perhaps more likely, an encoded probe sent by a civilization interested only in one-way communication. In any case, Loeb's point about being circumspect before deciding whether or not to reply to a strong signal is worth considering if we value the future of humanity on Earth.

Not everyone agrees, however. Russian astronomer Alexander Zaitsev, chief scientist of the Russian Academy of Sciences' Institute of Radio Engineering and Electronics, has already sent several powerful interstellar missives to a handful of sunlike stars using the Evaptoria Deep Space Center in Ukraine, one of the most powerful radio transmitters on Earth. Zaitsev believes that as a species we are morally obligated to tell other potential civilizations that they are not alone. This approach, known as Active SETI (or METI, Messages to Extraterrestrial Intelligence), has stirred much subdued controversy among SETI supporters and scientists, many who think if such broadcasts are to be pursued, they should at least be discussed first. John Billingham, a senior scientist at the SETI Institute, says, "We're talking about initiating communication with other civilizations, but we know nothing of their goals, capabilities, or intent. At the very least we ought to talk about it first, and not just SETI people. We have a responsibility to the future well-being and survival of humankind."[37]

An editorial published in the science journal *Nature* in 2006 echoed Billingham: "It is not obvious that all extraterrestrial civilizations will be benign, or that contact with even a benign one would not have serious repercussions." The editorial was written after a SETI study group of the International Academy of Astro-

---

*Remember that back in 1948, Arthur C. Clarke was virtually alone in his conviction that satellite communications would one day become a reality.

nautics voted against a process whereby the content of a reply to an ET signal would be deliberated in an open forum. In effect, the editorial argued, this decision meant that "anyone with a big enough dish can appoint themselves ambassador for Earth."[38]

But Seth Shostak, senior astronomer for the SETI Institute and current chairman of that study group, thinks that even if global guidelines were made, they would be totally unenforceable. "What are you going to do? Stop people from aiming their backyard antennas into space? Anyway, it's not clear to me that it matters terribly much because in a sense we've already broadcast our presence into space. We answered 'them' fifty years ago."[39]

Shostak refers to the electromagnetic noise of our radio and television broadcasts. As mentioned earlier, these signals have been spreading out from Earth in all directions for over a century. With distance, of course, they weaken considerably, so it would take a radio detector more powerful than any existing on Earth to pick them up from nearby stars. But if a civilization is 1,000 years ahead of us, is it not conceivable that its receivers may, in fact, be that powerful?

There is yet another way that Earth might stick out among the distant stars: the chemical composition of our atmosphere. Astronomers are on the verge of being able to spectroscopically analyze the atmospheres of extrasolar planets—or "exoplanets"—by looking for biological signs of life, or biosignatures. Life is like a machine in that it breaks down organic compounds to extract energy. This process, called metabolism, is carried out by us humans when we metabolize food, which, in turn, builds new cells and tissues, generates heat, and enables our bodies to perform physical tasks. Life, in whatever form it takes, thus creates biosignatures by interacting with its environment. The most prominent biosignature is oxygen, followed by methane and nitrous oxide. Telescopes already on the drawing boards, like the James Webb Space Telescope, will be able to study the immediate environs of stars at infrared wavelengths and detect the spectral signatures of any attendant planets, including biosignatures and, perhaps, "technosignatures" like the Earth's rising carbon dioxide levels.

If we are close to accomplishing this, then it is conceivable that an extraterrestrial civilization advanced by even a century and located somewhere in our corner of the Milky Way may have already detected Earth's own biosignatures and technosignatures. Our very breath may have given us away. Perhaps a signal from this civilization is even now streaking at light-speed in our direction, or perhaps ETs, or their robotic probes, are venturing toward our planet in hopes of investigating what their scientists may suspect is a young, industrialized civilization. If so, let's hope they bring with them the best of intentions.

While all of this may sound like so much fiction, we cannot deny that scientists are actively searching for signs of extraterrestrial life, that SETI is a branch of astronomy. That fact alone implies some hope of eventual contact with whomever *they* may be. Although largely an academic exercise for now, it is only natural for us to conceptualize the kinds of civilizations that may exist in the universe, particularly since such musings can lead us in productive directions concerning how we envision our own future. In broad-brush terms, such questions have already been considered by a host of philosophers, scientists, science fiction writers, deep thinkers, and just everyday folk. But for all these rarefied ruminations, at the end of the day we still end up with assumptions, which are all we're going to have until we get a few data points, such as a definitive signal from "them" or some other form of contact, like an invasion.

Some assumptions are better than others in that they are probably givens. For example, we can assume that, like us, ET has evolved in a region of the Galaxy that satisfies all the terms in the Drake equation: stable star, stable galactic region, a life-friendly planet, no life-erasing supernovae or gamma-ray bursts, no total mass extinctions, and certainly no total self-annihilations. Moreover, just as we have experienced in our neck of the cosmic woods, ET must also have ridden out a few ice age epochs, asteroid impacts, and other disastrous natural events beyond his control. On the other hand, these ETs could have been splitting the atom while we were discovering fire and, assuming they survived their nuclear growing pains, they may have advanced to the point where they are no longer vulnerable, or as vulnerable, to whatever the universe throws at them.

Just how far can a civilization go in controlling its part of the universe? According to a theory by Russian astrophysicist Nikolai Kardashev, truly advanced civilizations can go a long way indeed. In the early 1960s, he divided potential extraterrestrial civilizations into three main types, based on the amount of energy they could muster for interstellar communications.[40] Type I has evolved to the point where it can master planetary energy. Type II civilizations can master stellar energy forms; Type III, galactic energy. Types II and III civilizations, then, if threatened by an asteroid or a potential supernova, may be able to do something about it by calling on the kinds of energy reserves contained in stars and galaxies. Type I civilizations might be able to thwart, at the very least, an impending asteroid impact, but obviously not as effectively as the more advanced civilizations. Our civilization is probably just now entering the lowest regime of the Type I rank, considering our reliance on oil and coal, supplemented with a smattering of nuclear fission and alternative energy sources like wind and solar power. When you think about all the energy that we could potentially harness from just the Sun, our dependence on fossil fuels is like living off the crumbs dropped from a sumptuous banquet table at which we could be seated if we put our minds to it. But that's a topic for another day.

Whatever the ranks other civilizations may have in our vicinity of the Milky Way, SETI scientists are beginning to sense that some sort of detection may be in the wind, perhaps within the next couple of decades as the search technology evolves and globally propagates.[41] Although detecting an ET signal doesn't carry the visceral impact of a Wells-type of in-your-face alien invasion, it is nonetheless a cosmic knockout because, whether we think about such matters a little or a lot, it would significantly alter our perspective about our place in the universe. The big question, of course, is exactly how would our view of ourselves change?

No one is really certain, but there is no lack of intriguing, and sometimes mind-blowing, speculation. Richard Dawkins, in his book *The God Delusion*, writes that one possible reaction to getting our first communiqué from ET might be "something akin to worship" because any civilization capable of communicating over the vast distances of interstellar space would have to be greatly superior to our own. For a modern analogue, Dawkins cites Sir David Attenborough's expeditions to the South Pacific in the late 1950s, where he learned of the emergence of "cargo cults," which sprang up in Pacific Melanesia and New Guinea after the Second World War. When white immigrants began settling there during the first half of the twentieth century, the indigenous islanders, as unexposed to the modern world as Neolithic people would have been to something as simple as a flashlight, so marveled at the ceaseless procession of, what was to them, astonishing cargo that followed in the immigrants' wake (e.g., radios, tinned goods, electric generators, clothes with bright buttons, etc.), that they endowed these things with supernatural origin. Many cargo cult religions were subsequently formed, nearly all of them claiming that a "messiah" would bring them equally spectacular "cargo" on the day of their apocalypse.[42]

But even if we deem our extraterrestrial callers to be roughly on the same technological rung of evolution as ourselves *at the time they transmitted their message*, we would still have to accede that they are the more advanced race *by the time we receive it.* Why? Because even at the speed of light, the one-way travel time between the stars, depending on the distance, would take hundreds or even thousands of years. In other words, if we detected a signal tomorrow from a star that is, say, 200 light-years away, that would mean they first transmitted that message 200 years ago and are, hence, at least that many years ahead of us—technologically speaking—*now.* We could only try to imagine what kind of cargo would be off-loaded from their starships![43]

But we're getting a little ahead of ourselves here, insofar as speculation is concerned. For the moment, let's keep the ET intervention a simple one. How would society react to the detection of a simple modulated signal that was interpreted, and confirmed to be, from an extraterrestrial civilization? The very first bit of information the signal would give us would be the realization that we are not alone in the universe. There may be many who already believe this, but with the detection of an ET signal that assumption would become a certainty. Such news would be sure to have some immediate, galvanizing effect on the public. The SETI Institute's Shostak reminds us of NASA's momentous announcement in August 1996 that a meteorite from Mars appeared to contain fossilized microbes. The news sent shock waves through the science community, and, at least up to the point that the findings became controversial (and even questionable), the public interest was insatiable. He thinks something similar would happen with the detection of an ET signal. "Once it became clear that there was no immediate danger, that the aliens weren't about to land," says Shostak, "I think the short-term effect would be curiosity. People would want to know everything about them."[44]

This phase would last awhile, no doubt, but after the immediate flash of sensation, it is likely that public interest would fade as the weeks turned to months then years, especially if no further signals were detected. This kind of falloff is typical of media audiences who become saturated with a single news event, no matter how spectacular, and want no more of it. On the other hand, some scientists predict that there would be a slow infiltration of this new knowledge—that intelligent life exists elsewhere in the universe—into the deeper structure of society.[45] As time goes by, it would simply become part of our history and cultural makeup, in the same way as landing on the Moon has become another facet of who we are as a civilization (as in, "If we can put a man on the Moon, how come we can't end poverty?"). Exactly how this would manifest itself over time is up for debate, but it would permeate practically every aspect of society; society would have to adjust, sometimes a little, sometimes a lot. "A detection would justify the personal philosophies of some people," says Shostak, "while other people would probably adjust their personal philosophies accordingly."

One institution that might have to make the biggest adjustment of all is religion. Some theologians are already addressing potential issues in a field they call "exotheology," or the nature of God and religious belief as it pertains to extraterrestrial intelligence. The reasoning in some quarters sometimes suffers from reliance on a literal translation of biblical scripture or the doctrine of a particular denomination, but there at least appears to be greater dialogue on the subject today than in years past. Some hold that the discovery of extraterrestrial intelligence would not affect religion very much, no more than Galileo's discoveries affected Catholicism. For those who believe in an infinite God, the discovery would, in effect, illustrate the greater glory of God; in other words, "God can make us in this part of the universe and he can make intelligent anthropoidal newt-like creatures, if he so desires, in another part of the universe."[46] After all, they, too, could believe that God created the whole universe and no place or life-form is more special than another.

Other thinkers, however, foresee contact with ET as shaking religious thought to its core. In his book *Are We Alone?* physicist Paul Davies predicts that the new reality would force theologians to reappraise their religious doctrines, if not scrap them altogether. One of the major difficulties, he says, would be the assumption that we may be dealing with beings much more advanced than ourselves, beings whose religious beliefs may appear as advanced and inscrutable to us as Captain Cook's religious and scientific beliefs appeared to the South Sea Islanders and Aboriginal people he encountered in the eighteenth century. "The difficulty this presents to the Christian religion is that if God works through the historical process, and if mankind is not unique to his intentions, then God's progress and purposes will be far more advanced on some other planets than they are on Earth.... It is a sobering fact that we would be at a stage of 'spiritual' development very inferior to that of almost all of our intelligent alien neighbors."[47]

The day after an ET detection, Christian religions with the most fanatical dogmas might react the strongest (if they chose to act at all). One of the most formidable challenges they would have to face would be how to interpret their position of Jesus Christ as God incarnate in light of a universe potentially abounding with intelligent life-forms. Scholars who seize on this point contend that Christians would have to decide whether the incarnation occurred only on Earth or if God ordained it on a multitude of worlds brimming with a multitude of life-forms. If incarnation transpired only on Earth, what of the beings on other worlds? Would they be capable of redemption, though they may look like jellyfish and have no religious beliefs per se? Some religious scholars who take the Bible literally argue that such questions are irrelevant because the Bible doesn't address the existence, much less the salvation, of ETs. "Scripture," they point out, "strongly implies that no intelligent life exists elsewhere.... Earth was created to be home for creatures made in God the Creator's image."[48]

Of course, for those who believe the Bible is the end of the story in such matters, there can be no other truth and certainly no external correction from science. If the aliens contact us one of these days, their signals won't amount to anything more for those believers than the fossils God "planted" on Earth 6,000 years ago for our amusement. But for those who have retained both their intellect and their faith, the speculation spreading from that single seed of a signal will no doubt grow like a wild vine. Even Osama bin Laden, sitting in his cave, would probably wonder who these new alien infidels were. In fact, they could be anybody with their own morals, belief systems, and creation myths—extensions of cultures on Earth today taken to extreme development. For example, the aliens might reflect nontheistic Buddhist beliefs, embrace the cosmogony of the Navajo or Pueblo, or have no concept of God at all. What then? More interesting still, what if the ETs try to foist their own "bible" on us, much as the Spanish Catholic missionaries did to indigenous Native Americans of the southwestern United States in the early seventeenth century? Whose bible would win?

It doesn't take a sociologist to conclude that, beyond the "we-are-not-alone" realization, detecting a signal from ET could have numerous, unforeseen long-term repercussions. Although it might not hit civilization's reset button like that of an asteroid impact, a successful detection would, in the words of Jill Tarter of the SETI Institute, "calibrate our place in the cosmos and provide a mirror with which we can see humanity as all the same when compared with another independently evolved form of technology users."[49] Some serious hypothetical questions would have to be asked, with many necessarily probing the weakest points of some of our most profound beliefs, if not knocking the legs out from under them. Depending on the individual, emotions would likely run the gamut from childlike wonder to existential dread. Yes, contact, if ever it occurs, will profoundly alter the defining mood of the world in countless unforeseen ways.

Naturally scientists, being scientists, hope that there won't be just one signal, but a stream of information from the advanced civilization, passed along in stages so there would be time for us to assimilate the new knowledge. Imagine getting information spoon-fed by a race of beings thousands or millions of years ahead of us! What would this astonishing new knowledge contain? Perhaps new insights into the nature of the universe, plans for constructing a faster-than-light spacecraft, ways of harnessing energy from the stars, ways to teleport matter, a plan for how to build a traversable wormhole as in the film *Contact*, or (assuming a similar biology) a plan for how to eliminate disease. The information could be as sensible as ways of achieving a more universal understanding when it comes to conflicts and problem solving, or as radical as the means of surviving "far beyond our own level of evolution."[50]

But we're assuming ETs would want to foster further contact, perhaps even strike up a rapport. That might not be the case. It's just as possible that, other than noting our whereabouts in the Galaxy, they may have little interest in us or prefer a hands-off approach based on a noninterference principle, similar to *Star Trek*'s Prime Directive. That would certainly be humbling to us but much preferable to a physical confrontation with ETs or their robots, which could lead to a neo-Darwinian-like domination of our culture by theirs and the end of civilization as we know it.[51]

Another eventuality could present itself. What if, after another half century or so of searching, no extraterrestrial civilization steps up and rings our doorbell? How, then, do we explain the silence? There are any number of possibilities or combinations thereof, including that we truly are alone in the universe or, at best, that ETs are exceedingly rare.* If the latter is the case, then ETs may be in a different developmental state—inventing stone tools rather than vacuum tubes or evolving to the extent that primitive civilizations such as our own no longer interest them. They may even be extinct.

It is here that we encounter the Achilles' heel of any endeavor to predict the probability of alien life in the Galaxy. Even if we knew with any certainty how long a typical civilization communicated by using radio waves, for example, we can't ignore the Drake equation. Other factors would need to be considered, such as the number of stars that can harbor planets, the number of planets that can harbor life, and the kind of life that evolves to sentience and that does not destroy itself but instead employs radio telescopes to listen to the stars. You can constrain these variables all you want, change the order around and modify each as you see fit, but the final product will always be at best an educated guess, with a story line that is more about "us" than about "them."

---

*This is related to the "where are they?" argument, first proposed by Enrico Fermi in 1950: Earth is much younger than the universe, so if alien civilizations arose billions of years ago, they should have colonized or at least visited Earth by now. There is no evidence of either. Ergo, we are alone.

Nevertheless, even though not one salient ET twitter has been heard in over half a century of listening, SETI astronomers (who must be the most patient, optimistic, glass-half-full people in the world) remain undaunted. "We are a very young technology in a very old Galaxy," reminds Jill Tarter. "Fifty years only means something when referenced to human rather than cosmic time." Tarter has been in the SETI business for decades and knows very well that researchers have their work cut out for them, but she also firmly believes that the search is one of fundamental importance to humanity. She remains philosophical about the SETI endgame. "In the past 50 years we haven't searched even the smallest piece of the cosmic haystack. The searches will either succeed and spawn new searches, or they will cease when some societal threshold of pain gets passed and we are willing to accept the incredibly important negative, rather than pay for more searching."[52]

What a sad day for the human race, if it ever comes to that. We would be turning our backs on one of the oldest questions that has ever been asked: Are we alone? What would make abandoning SETI all the more absurd is how much closer we are today to really finding out an answer to that question. Astronomers once wondered if planets orbited other stars or if the solar system was somehow unique. Now they know. Other planets—other solar systems—*do* exist. There are more than 300 known extrasolar planets, and scientists estimate there could be as many as 30 billion planets in our Galaxy alone. So far, most of the known exoplanetary systems have planets with Jupiter-like masses, but astronomers are beginning to find smaller ones with planets spaced like our own solar system. At least ten have masses that fall between 5 and 20 times that of Earth, qualifying them as "super earths"—a completely different, and intriguing, variety among known planets.[53]

Up to now, we've found only large planets because large planets are easier to find with current planet-searching capabilities. But telescopes are improving and methodologies are being refined and new ones invented. Next year, or the year after, the list of super earths will grow, and eventually, perhaps very soon, the first earthlike world will be discovered. The star harboring that world, wherever it is, will be the most sought out in the night sky for years to come. It won't change the universe, but it will most definitely change the way we look at the universe and ourselves, because where there's one blue world there are very likely many, and the more blue worlds there are out there, the greater the chance that we are not alone.

Unless the human race totally degenerates in the future, the search for extraterrestrial intelligence will likely endure as long as civilization itself, whether ETs are never found or found in scores. SETI is an experiment that cannot fail, because the search remains one of the few ventures that encourages us humans to glance up from our plate of concerns on Earth and acknowledge, if only for a moment, our cosmic roots among the stars. If intelligent "others" do exist in this universe, then they, too, must have come from the stars. Our similarities, however, may end there, and that is something we may have to live with.

CHAPTER 8

EXOTICA

"How—how does the Universe end?" said Billy.

"We blow it up, experimenting with new fuels for our flying saucers. A Tralfamadorian test pilot presses a start button, and the whole Universe disappears." So it goes.

—Kurt Vonnegut, *Slaughterhouse-Five*

A t noon on May 19, 1910, and for several hours thereafter, the sky in southern Wisconsin put on a spectacular display of color never before seen by sky watchers. Astronomer Edward Emerson Barnard of Yerkes Observatory recorded the peculiar atmospheric conditions, which were also observed by the staff and, indeed, in other parts of the world: "A horizontal bar of brilliant prismatic colors (red above) about 25° or 30° long, was visible in the south at an altitude of about 20°. This phenomenon was produced among high cirrus clouds. The low cumulus clouds from the south obscured the band in passing. Around the sun was a prismatic halo, 22° in diameter, as measured by Mr. [Oliver J.] Lee with a theodolite. All about the sun were patches of iridescence on the cirrus."[1]

Was it just a coincidence that at about this same time, Earth was passing through the dusty tail of Comet Halley? Astronomers the world over were reporting that the sky around this time had a milky, hazy appearance throughout the night. On the evening of the twentieth, the head of the comet was observed by English solar astronomer John Evershed, "with [its] short tail (evidently much foreshortened) directed in the normal way opposite to the Sun."[2] In fact, the tail at that time would have been directed toward Earth, and that's why it looked foreshortened. On the twenty-first, Evershed reported that the tail passed through the square of Pegasus, "which was filled with faint light." Other observers reported seeing "strong light to the N.N.E. under Cassiopeia" and the "band over the N.E. horizon looked exactly like a fine display of the Zodiacal light; it was of a pearly white colour."[3]

The apparition of periodic Comet Halley in 1910 was a close pass, though its nucleus—which really could have done some damage had it struck Earth—never got any closer than 60 million kilometers (37 million miles). The swept-back tail, however, was broad and long enough for Earth to penetrate with ease. Its passage would have been an exciting event enjoyed by all, save for the fact that astronomers had learned in years prior that comets contain cyanogen, a gas that can be deadly depending on one's exposure. Unfortunately, a few newspapers published sensational stories reporting on the supposed danger, and, despite astronomers' reassurances that there was nothing to worry about, a minor panic ensued. Anti-comet pills were hawked by the thousands while more imaginative shysters advised people to hire out submarines to keep them safely below the ocean until the cyanogen dissipated. Of course, the comet came and went with no adverse effects.

Of all the astronomical events that can happen, comets seem to bring out the worst in people. They are either blamed for unhappy events (the Black Plague, for instance); are considered omens of said events (the fall of Jerusalem in the year 70 BCE); or inadvertently trigger them (the mass suicide of 39 members of the Heaven's Gate religious group in California in 1997). Comets are easy to fixate upon because, though they are fairly common visitors to the inner solar system, the really bright ones are relatively rare. The beguiling apparitions virtually ensure that a host of crystal gazers and hooey mongers will pop up with bombastic interpretations for the gullible folk among us. If you don't believe me, type "comet prophecy" in the search box of your favorite search engine. You will see that the promulgation of false and alarming stories based on celestial portent is still alive and well in the twenty-first century.

Other than this kind of high-profile celestial event—to which meteor storms and lunar and solar eclipses may be added—most people don't give the stars a second glance. And why should they? Earth, along with the Sun, may be moving through interstellar space, but what happens "out there," though it may sometimes attract our attention, has always stayed out there. Like the desert, the weather in space tomorrow seems as predictable as it is today.

On human timescales, this blasé attitude is to be expected, but as I hope the reader by now appreciates, when cosmic timescales are factored in, the universe can intervene on planetary playing fields in untold ways, some subtle and some not so, and, as we shall see, a few of them in quite unique ways. For simplicity's sake, I've divided these events into three categories: treacherous encounters, game-over galaxy collisions, and sinister singularities. The events described in each category are all astrophysical possibilities, though some, such as a cataclysmic phase transition, might be termed "one-off events" because they obviously don't recur with the regularity of, say, supernovae or asteroid impacts—if, indeed, they recur at all over the history of the universe. But an important point to bear in mind throughout this chapter is that predicting the likelihood of highly unlikely cosmic events is made all the more problematical when considering how little we know about the nature of the universe. Although astronomers know that roughly 5 percent of the universe consists of normal, touchable, matter—that's you and me, trees, oceans, planets, and stars—they remain quite uncertain as to what makes up the remaining 95 percent. This state of affairs is like trying to understand why Earth is the way it is without giving any thought to the fact that three-quarters of its surface is covered in water. Such a void in the knowledge base leaves astronomers more than a little annoyed. Said noted cosmologist Sir Martin Rees, "It's embarrassing that 95 percent of the universe is unaccounted for."[4]

Does that also mean the laws of physics can describe only 5 percent of what can happen in the universe, or that the kind of astronomical events Earth has experienced over the course of its evolution amounts to only a pinch of what's possible? If so, there may be cosmic connections that remain not only to be seen, but imagined.

BAD NEIGHBORS

As detailed earlier, the solar system over its several-billion-year history has been exposed to a variety of interstellar environments. Typically these amounted to variations in the density and temperature of the immediate interstellar medium, producing conditions that could have altered Earth's climate. In addition, direct exposure to supernovae or gamma-ray radiation and their propagating shock fronts may have even played a role in modifying the evolution of life on the planet. Fortunately, the Sun is currently located in a relatively empty region of interstellar space, with an average density of 0.1 atoms per cubic centimeter. That's more vacuous than the best laboratory vacuum on Earth. However, though seemingly empty, the small-scale structure of the galactic environment of the Sun remains largely unexplored, and the possibility of the solar system encountering an uncharted "cloudlet" of interstellar dust and gas, or even an overlooked, free-floating "brown dwarf" star, cannot be ruled out over the next tens of thousand of years. The shadows beyond the yellow light of the Sun may yet be deep enough to hide small, but hazardous, secrets in their folds.

It's even possible that we might not completely understand how space and gravity work in the interstellar medium beyond the Sun. How so? Consider the mysterious force acting on the Pioneer 10 and 11 probes, the ones that flew by Jupiter and Saturn back in the 1970s, never to return. Even though they have since been barreling out of the solar system in opposite directions at about 12 kilometers (7.4 miles) per second and are now over 12 billion kilometers (7.5 billion miles) from the Sun, precise measurements made of their relative motions with respect to the Sun over the years have revealed that something strange is happening to both probes: they are not where the conventional laws of physics predict they should be. Both spacecraft, after over three decades of travel, should be about 400,000 kilometers (250,000 miles) farther from the Sun than they are, but it seems as if something is either trying to pull, or push, them back. (Astronomers refer to this motion as an "anomalous acceleration toward the Sun.") Each year, their forward momentum is reduced by about 5,000 kilometers (3,000 miles).[5] What could be causing this?

Since the discovery of the "Pioneer anomaly" in 1980, scientists have become increasingly more desperate to explain it. At first they proposed conventional mechanisms, such as drag caused by comet dust, gravitational perturbations by small clouds in the outer solar system, or thermal recoil from heat emitted by the spacecrafts' own batteries and electrical systems. More recently, they've resorted to more exotic possibilities, including dark matter (which might somehow be throttling the probes' forward motion), some heretofore unknown effect of gravity over great distances, and even an inconstancy in the speed of light. So far, however, the origin of the probes' deceleration remains an intriguing unknown.

Of course, a perfectly commonplace explanation for the Pioneer anomaly may be forthcoming—or maybe not. If astronomers were completely sanguine about what was going on with these probes, they probably wouldn't be proposing, among other things, modifications of Newtonian dynamics or Einstein's theories of relativity. Whatever the explanation turns out to be, the point is we're just beginning to study an environment that is billions of years old, and not every surprise can be anticipated. About all we can do is to continue looking for new clouds and new stars in the neighborhood so we can become better aware of our immediate celestial surroundings.

Great strides toward this are being made all the time. Astronomers now know of at least two, and maybe four, interstellar clouds looming on our cosmic horizon. One, called the G Cloud, lies less than 5 light-years away; another, called the Apex Cloud, is a little over 16 light-years away. By plotting the Sun's motion with respect to these clouds, astronomers have determined that the Sun should enter the G Cloud in about 45,000 years, give or take, and the Apex Cloud in 175,000 years.[6] In addition, there may be two interstellar cloud fragments lying in the gulf between the

Sun and Alpha Centauri, the Sun's nearest neighbor, about 4 light-years distant. Spectroscopic observations show that the light coming from Alpha Centauri is being absorbed by some intervening gas. The absorber could be a dense cloud or nothing more than a little turbulence in the interstellar gas. In either case, these should sweep past the Sun within the next 3,000 years.[7] The repercussions of the encounters with the G and Apex clouds are unclear, but their densities may be enough to dramatically affect the Sun's galactic environment, as well as the physical properties of Earth's magnetosphere and atmosphere. As for the clouds drifting between the Sun and Alpha Centauri, until further observations can be made, their consequences remain yet another intriguing unknown.

I should mention here that the other potentially unpleasant aspect of tangling with an interstellar cloudlet is that it could disrupt the swarm of comets arrayed in the outermost reaches of the solar system, a region called the Oort Cloud some trillions of kilometers beyond the orbit of Pluto. The gravitational wake of such an encounter could send millions of kilometer-size iceballs toward the Sun over periods of a few million years. But interstellar cloudlets are not the only objects that can dislodge comets from their maundering orbits. Other stars passing near the Sun could do the job just as well. It turns out there are well over 200 stars within 33 light-years of the Sun, and more are being discovered every year.[8] Thanks to stellar surveys like the European Space Agency's Hipparcos mission, the velocities and motions of many of these and other stars in the solar neighborhood have been reckoned down to sub-arcsecond tolerances, allowing astronomers to predict when, and how close, a star will approach the Sun over periods of millions of years. Fortunately for us, encounters are, and will remain, rare.

The most notable future case is a minor red dwarf called Gliese 710, located some 60 light-years away in the constellation Serpens. Gl 710 will pass within a light-year of the Sun in 1.4 million years, close enough to impinge on the Oort Cloud. The prediction is that Gl 710 could increase the number of comets with Earth-crossing orbits by 50 percent over a period of 2 million years. Many will make subsequent return visits to the inner solar system, similar to Comet Halley's repeated sojourns to the Sun, but about half will be ejected on their first passage. At any rate, Gl 710 will likely create a minor comet shower, as brushes with other stars have done countless times in the past.[9]

If you stop to think about it, the only reason we know about the existence of comets is because at some point in the dim past they were diverted from their far-flung orbits in the Oort Cloud toward the inner solar system by passing stars and interstellar clouds. The idea of comet bombardments, however, is more popularly associated with the hypothesized existence of a low-mass companion star to the Sun, described earlier, called Nemesis. But while the existence of Nemesis is still debated (and pretty much discounted), the fact remains that the solar neighborhood is a more dynamic and protean environment than once believed, and hence a disturbance by either a small interstellar density enhancement or a stunted star is not only plausible—it is inevitable. A 1999 study found that some comets in the outer Oort Cloud have unusual orbital motions that can best be explained by "something" (the researchers suggest a brown dwarf of about 3 Jupiter masses) coming along and giving them a gravitational kick at a distance of nearly 4 trillion kilometers (2.4 trillion miles) from the Sun.[10] Whether or not such a low-mass solar companion exists, the bottom-line message brought to us by comets from the Oort Cloud is that the solar system is connected to, and affected by, the rest of the Galaxy. The next time you see a comet in the sky, thank a passing star or an interstellar cloud for sending it our way.

In addition to stars, putative stellar companions, and interstellar clouds, there are still other kinds of interstellar exotica that the Sun could conceivably encounter in its journey around the Galaxy. Recently, astronomers put forth the theory that the first stars to form in the universe were not brilliant suns but invisible "dark stars." These objects would have been powered as matter met its antimatter counterparts and annihilated them—specifically, the lightest of the dark matter candidate particles known as the Lightest Supersymmetric Particle, or LSP. According to this theory, as the protostellar halo would try to contract into a conventional star, the LSP annihilations would produce quarks and antiquarks that, in turn, would fluff up the "star," keeping it hot and voluminous. This process would also prevent the fusion of hydrogen into helium at its core and, as such, the production of heavier elements (which would be produced by normal stars). The final product, then, would be a huge, dusky starlike object that, if it still existed, would contain unprocessed primordial hydrogen.[11] As no dark stars have ever been observed, however, their evolutionary path—indeed their very existence—remains open-ended. No one can say whether they lived and died during the first billion years of the universe or are still around today.

If they are still around, as some astronomers speculate, they could be detected by their emission of copious gamma rays, neutrinos, and antimatter. Moreover, they would be monstrous in size, anywhere between 1 billion and 300 billion kilometers (620 million to 186 billion miles) across. Exposure to the environment of a dark star would be lethal to any life-bearing world. Hopefully, none are lurking near the Sun, although, admittedly, the detection of such objects, whether in our Galaxy or in the distant universe, would have profound implications for stellar evolution and, perhaps, life itself. For one thing, dark matter, the mysterious stuff said to make up about 25 percent of the universe, usually figures into theories about the collapse of gas clouds into the first galaxies in the early universe, but it has never been considered in the formation of the first stars. Hence, their possible existence could shed some light (so to speak) on the nature of dark matter as well as how massive black holes formed so quickly in the very early universe. A dark star, at a safe distance, would be quite a find.

GALACTIC SMASH-UPS

If you look at images taken of galaxies in the early universe, a billion or so years after the big bang, it is obvious that they are not the placid, whirlpool-like objects we see in the nearby universe. Many resemble shards of broken glass, recognizable as something that could be called spindly or spherical or vaguely shaped like a gull's wing or a broken seashell, but more than that you cannot say. They look as windblown and random as confetti. But galaxies they are. The reason many of them look the way they do is because they've endured collisions, side swipes, and near passes with other galaxies, intergalactic clouds, and star streams. The history of these galactic brawls are thus reflected in their irregular shapes. The Milky Way, as unperturbed as it may appear, has itself collided or interacted with other galaxies in its long history. One day, in less than two billion years—still well within the lifetime of the Sun and, perhaps, the lifetimes of future astronomers (wherever they may be in the Milky Way)—it will do so again, not with a strange galaxy, but with its old neighbor the Andromeda Galaxy.

Being only 2.5 million light-years away, you'd think the Andromeda Galaxy would be an impressive sight in the sky, but it's not. It can just be seen in mid-November evenings with the naked eye in the constellation Andromeda as a diaphanous patch of oval light high in the northern sky. To find it in the Northern Hemisphere, all you have to do is go out around 8 o'clock, face south, and look almost straight up. Your sky has to be fairly dark and either moonless or nearly so. In the deep Southern Hemisphere (Sydney, Australia, or Cape Town, South Africa), you'll have to face north and look about 15 to 20 degrees above the northern horizon (less than a third of the way up in the sky). Binoculars are a great help; in fact, the galaxy looks best using low magnification because too-high magnification washes it out.

At any rate, that patch of barely perceptible light is steadily moving toward the Milky Way at about 120 kilometers (75 miles) per second, or 7,200 kilometers (4,500 miles) per hour. It hardly seems a threat now, particularly as most people don't even know it can be seen from their suburban backyards. But if you could somehow fix your gaze in this direction for the next two billion years and watch as the millennia flashed by like seconds, you'd see that diffuse patch of light grow larger and larger on the sky. At first, it would appear to be essentially a larger version of what we see now: a disk inclined only 12° from edge on with a bright milky center. About 800 million years from now, at a distance of about a million light-years, the long axis of the galaxy's disk will be at least 8° in length and nearly 3° across on the sky, larger, at least in length, than the Large Magellanic Cloud in the southern sky. Some 500 million years after that, it will be over 16° in length, which is roughly the length of the handle of the Big Dipper. By then, however, the massive dark-matter halos surrounding both galaxies will be in contact. This, along with the gravitational tidal effects of the two galaxies, will begin distorting their shapes to the extent that, from the Milky Way, the Andromeda Galaxy will no longer look disk-shaped but rather distended (as will the Milky Way as seen by astronomers located in the Andromeda Galaxy). Once it reaches a distance of less than 320,000 light-years, the galaxy will be stretched in all directions, filling the sky with a powdery bloom of light punctuated by bright star-forming regions. The galaxy's center, if it can still be seen from Earth, will appear as a broad, softly glowing orb shot through with stars.

The collision, when it occurs, will be a gentle event lasting hundreds of millions of years, with the two galaxies passing into and out of each other like converging smoke rings. A second passage will occur more quickly, about 1.5 billion years later (or 3.5 billion years from now), this one hurling long, curving arcs of stars hundreds of thousands of light-years into extragalactic space. The grand finale will occur 5 billion years from now when both galaxies fall back onto one another, merging finally into a huge spherically shaped entity of several hundred billion stars. By then the Sun will be little more than a cinder, and future astronomers will have to station themselves on other planets. But even then it is likely that their sun, wherever it is located in the Milky Way (or Andromeda), will experience some sort of disruption. Their journey through the new galaxy will be one wild ride indeed.

The inevitability of a collision between the Andromeda and the Milky Way galaxies was explored in 1959, when astronomers, using Kepler's tried-and-true laws of motion, determined that both galaxies had already interacted once, tracing out almost a full period of their orbital motion, and were in the process of falling back together.[12] The calculation at the time was that the next close passage would occur in about 4 billion years. But those calculations were made in the days before it was realized that all galaxies are immersed in extended, massive dark-matter haloes, which, in the case of the Milky Way and the Andromeda galaxies, exert more than twenty times the mass of the visible matter. The added mass exerts significant dynamical friction on the two systems, soaking up orbital energy and angular momentum, thereby speeding up the merger process by 2 billion years.

Despite the fact that both galaxies contain hundreds of billions of stars, few if any are expected to barrel into each other during the merger. When compared to the diameters of stars, the distances between them are immense. It's the same reason collisions between stars in the Milky Way or in any other galaxy are unlikely. On the other hand, countless numbers of stars will be perturbed by the gravitational effects of the collision, and some will even be slung from one galaxy into the other. Images of merging galaxies present a messy visage that results in plumes and shells of shocked gas, and "tails" of stars stretching away from the main disks of the galaxies like the splayed arms of someone tumbling out of control down a steep hill.

The question, then, is what will happen to the Sun? Obviously, by the time the two galaxies have merged into a single entity 5 billion years from now, the Sun will be a hot white dwarf and Earth an orbiting ash pile. But before that, during the first and second encounters, the Sun will still be around, and so, perhaps, may be the Earth. To find out, astronomers Thomas Cox and Abraham Loeb of the Harvard-Smithsonian Center for Astrophysics in Cambridge, Massachusetts, used a powerful numeric computer program* that is normally employed for cosmological simulations to model the future encounter to an unprecedented degree.[13] Their results were intriguing, to say the least. On the first passage,

---

*More about this program, called GADGET-2, can be found at this Web site: http://www.mpa -garching.mpg.de/gadget/.

they calculate a 12 percent chance that the Sun will be pulled from its present position, 26,000 light-years from the center of the Milky Way, and drawn into an extended tidal tail of stars 65,000 light-years from the former center. That chance goes up to 30 percent after the second close encounter. By the time of the final merger, there is a 68 percent chance that the Sun will have been ejected to some location in the outer realms of the new galaxy. Cox and Loeb also say there is a slight chance, only 2.7 percent, that Andromeda will pluck the Sun from the Milky Way during one of the passes, but since both galaxies eventually merge into a single system anyway, it hardly seems to matter.

Cox and Loeb also predict that, since the star-formation rate in the Local Group of galaxies has been in a steady decline throughout its evolution, and the gas content of both the Andromeda and Milky Way galaxies is already low, the merger will result in only a modest burst of star formation. Moreover, by the time the actual merger occurs, most of the gas in the galaxies will have been consumed by low-level star formation, further robbing the galaxies of what little star-making fuel is left. The final remnant, which Cox and Loeb have whimsically christened Milkomeda, will, indeed, be a milquetoast of a galaxy, spheroidal in shape, similar to an elliptical galaxy but much smaller and of lower luminosity. The inner regions will be more diffuse than present-day ellipticals and populated with old stars tracing out a wide range of orbits and orbital velocities, like moths flitting about a yellow streetlamp. And because it will be a spheroid, any astronomers observing from within Milkomeda at that time will see primarily the stars of their own galaxy, much as an observer posited within a globular cluster would not be able to see beyond the stars of that cluster. Moreover, if the universe continues to accelerate in its expansion, Cox and Loeb predict that extragalactic astronomy will come to an end within 100 billion years as the last visible members of the Local Group recede beyond our event horizon. In effect, the universe will disconnect itself from whomever, or whatever, still resides in any galaxy anywhere. The final merger of the Milky Way and the Andromeda galaxies will signal the beginning of the long cosmic dénouement toward cosmic isolation.

Well, that *is* 100 billion years from now, a great deal of time to wait for even the rarest of events in this book to occur. But what about in 30 million years, less than one-eighth the Sun's orbit around the Milky Way? That's when a giant comet-shaped cloud of hydrogen gas is expected to collide with the Milky Way's outer spiral arms. This extragalactic interloper, known as Smith's Cloud, measures 11,000 by 2,500 light-years and is falling toward the Milky Way at a speed of 240 kilometers (150 miles) per second.

Currently, it lies 8,000 light-years away, but astronomers say that the cloud's leading edge is already interacting with gas from our Galaxy, so one could argue the collision is occurring now.[14] The origin of the invading cloud is unknown. It may have been stripped from another galaxy that passed too close to the Milky Way or it may be a bit of leftover fluff from our own star system. In any case, it is definitely bound to the Milky Way now, and if it continues on its current trajectory it will fully fall somewhere between the Perseus arm and the next inner arm, the Carina-Sagittarius arm, about 90° ahead of the Sun in the disk. When that happens, it will likely trigger a burst of star formation. Many of these stars will be massive and, a few million years later, will explode as supernovae. The supernovae will be too far away to endanger life on Earth, but any established or developing life-forms in that sector of the Galaxy had better watch out.

No other extragalactic clouds or galaxies are known to be

Figure 15. A giant cloud of hydrogen gas, called Smith's Cloud, is plummeting into the outskirts of the Milky Way. The cloud's core will strike an outer galactic arm between 20 million and 40 million years from now. The cloud (lower right in this artist's conception) is an actual image made with the Robert C. Byrd Green Bank Telescope. This image was generated with data from telescopes of the National Radio Astronomy Observatory, a National Science Foundation Facility, managed by Associated Universities, Inc.

headed toward a calamitous rendezvous with the Milky Way, but it's likely that there are others lurking about. Such interactions may be more common than we think. In the past, anyway, interactions with material outside the disk of our Galaxy have been known to produce rather significant effects. An encounter, or encounters, with other small galaxies or massive clouds may explain why the Milky Way's disk is warped out of the plane by some 9,000 light-years. Some astronomers think the objects that so perturbed the Galaxy's disk may have been the Large and Small Magellanic Clouds. However, this idea has been thrown into question because of the mystery of the Magellanic Stream, a stringy band of neutral hydrogen gas that stretches for over 110° of sky toward our Galaxy from the pair. The stream, which contains between 100 million to 1 billion solar masses of gas, has long been thought to have been drawn out of one or both galaxies during a close pass of the Milky Way some 200 million years ago. But recent measurements of the three-dimensional velocity of these galaxies indicate that they are moving twice as fast astronomers thought. (The Large Cloud alone is moving some 400 kilometers [250 miles] per second, which is close to the escape velocity of the Milky Way.) As such, the Magellanic Clouds may not be gravitationally bound to the Milky Way at all but just making their first, and perhaps only, passage through the neighborhood.[15] If true, the Magellanic stream wouldn't have been tidally stripped and it wouldn't have created a "wake" through the Milky Way's halo. Rather, the stream may simply comprise material ejected from the Magellanic Clouds by star formation and supernovae and left behind in intergalactic space, like cigarette smoke trailing out of the window of a moving car. This excess velocity also means that an object with the mass of the Large Magellanic Cloud would have been moving too fast to have produced the warp in the Milky Way's disk. It seems as if the textbooks on this matter may have to be revised.

DIMENSIONAL OBLIVION

No discussion of cosmic mayhem would be complete without mention of what might happen if the solar system encountered a black hole, or a black hole encountered the solar system. Such an occurrence is considered extremely improbable, and yet astronomers know of at least one such rogue black hole that is even now passing through the outermost reaches of the solar neighborhood. On March 29, 2000, the Rossi X-ray satellite detected X-ray outbursts coming from a small red star, cataloged XTE J1118+480 (the numbers reflect the star's positional coordinates in the sky). Follow-up observations revealed that it was orbiting an unseen companion once every 4 hours. Since no optical counterpart could be detected, astronomers concluded that the companion was, in fact, a black hole with a mass between six and seven times that of the Sun.[16] It was the eleventh such object known at the time. (As of this writing, there are 18 confirmed black hole binaries out of an estimated total population of 300 million stellar-size black holes believed to exist in the Galaxy.)[17] Observations using both radio and optical telescopes show that the object is a scaled-down version of a quasar, or a "microquasar." Just as with a normal quasar, material from the star is being drawn into a disk that wheels rapidly around the black hole's event horizon, the threshold that is the point of no return for any matter that passes through. The resulting friction heats up the gas to the point that it emits X-rays, but when a clump of material falls into the black hole, it emits a burst of brighter X-ray emission, a kind of farewell "banzai!" before passing into oblivion. At the same time, the black hole's magnetic field creates two jets of ionized gas that beam radio energy from the poles, telling the rest of the universe of its presence.

These qualities alone make XTE J118+480 a fascinating study, but in 2001 astronomers using the Very Long Baseline Array (VLBA) radio telescope were astonished to find that the pair is zipping through interstellar space at a speed of 145 kilometers

(90 miles) per second, about four times that of typical stars. The velocity was determined through a combination of the VLBA's ability to pinpoint the star's movement against the backdrop of more distant sources and by comparing its position in images taken 43 years apart for the Palomar Observatory Sky Survey. This also allowed astronomers to calculate the black hole's orbital path around the Galaxy. Its journey appears to have begun somewhere in the halo, 9,000 or so light-years above the galactic plane. Over the last 240 million years it passed through the disk three times, looping wildly around the center of the Galaxy until it entered our galactic quadrant some 37 million years ago. It only recently passed north of the Sun, missing the solar system by a generous 6,000 light-years.[18] Still, given that 6,000 light-years makes up only 6 percent of the Galaxy's total diameter, one might consider that distance a little too close for comfort.

What makes this little microquasar such a challenge to understand is not so much its very existence but its unusual orbital configuration and speed. Astronomers propose three theories (so far) as explanations. Perhaps XTE J118+480 was batted out of the Milky Way's disk when the black hole's progenitor exploded as a supernova; the small companion somehow survived the explosion and was simply taken along for the ride. But there's a problem: if the supernova explosion was symmetrical, it would have required the ejection of more than 40 solar masses, an improbable number, say astronomers. Alternatively, an asymmetrical explosion—one in which the impetus of the blast was directional—could have imparted an additional kick.[19] Astronomers think so-called runaway neutron stars got their unusual velocities from sideways, or directional, explosions. But there's a problem here as well. Though XTE J118+480 is only seven times more massive than the Sun, it is nonetheless far more massive than any neutron star, and to propel it to its current velocity would have required a kick of unprecedented force. Enter theory three, which supposes that the black hole's progenitor was an ancient and very massive star that formed in a globular cluster in the Milky Way's halo. Computer simulations of the gravitational interactions among the crowded stars of globular clusters show that black holes that form from the collapse of the cluster's most massive stars are eventually ejected, usually by mergers with other black holes. The kick can conceivably propel a newly merged black hole up to velocities as high as 4,000 kilometers (2,500 miles) per second.

And now we enter a new realm of speculation, because if this is true, then each globular cluster in the Galaxy could spit out many stellar- and intermediate-mass black holes into interstellar space. Given that there are at least 200 known globulars in the Milky Way, and probably many more unknown to us, there may be hundreds, even thousands, of black-hole vagabonds wandering invisibly through the Galaxy. Not only that, but astronomers know of many other astrophysical processes that can create such black holes. In truth, then, the Galaxy may be aswarm with as many as 100 billion black holes ranging in size between 30 and 300 kilometers (18 and 186 miles) in diameter. At worst, one of these tiny dimensional drains could swallow up our Sun or the Earth; at the very least, a near miss might disturb the Oort Cloud, sending a barrage of comets toward the inner solar system, whereupon the multiple impacts would wreak havoc on Earth and perhaps even destroy all life.

Relax! Astronomer Kelly Holley-Bockelmann of Vanderbilt University, who has made a study of these rogue black holes, reported at the January 2008 meeting of the American Astronomical Society that the odds of the Earth actually encountering one is on the order of one in a quadrillion. The universe is big;

black holes are small, even relative to the size of the Earth. We would have to be within a few hundred kilometers of the event horizon to be in danger of being sucked into oblivion.[20]

Should one lurk nearby, it would be difficult to detect, says Holley-Bockelmann, unless it had a companion like XTE J118+480 that was being visibly cannibalized. Such a spectacle would be obvious many light-years from Earth. If the encroaching black hole was isolated, however, about the only way to see it coming would be the way in which its intense gravitational field would bend the light of more distant stars along its line of sight, making them appear to shift anomalously in position and in brightness. The discovery of such an object, of course, would be of intense interest to astronomers, who would track the phenomenon for years after determining its orbit. As it drew nearer, it would no doubt interest the rest of humanity as well, for obvious reasons. A rogue black hole the size of XTE J118+480 could exert a gravitational presence for a few million kilometers around it and, much closer in, even twist the space-time continuum around in the direction of its rotation, a phenomenon called frame dragging. The effect is too small to notice for a rotating body like the Earth or even Jupiter, but for one as massive as a black hole, the space-time immediately surrounding it would be distorted into kind of a vortex. But as for sucking the Earth directly into its greedy maw, it would still have to pass very close. Given how small a target the Earth is, such a scenario is extremely unlikely, says Holley-Bockelmann, which is why this cosmic connection falls under the "exotica" category.

As does our next and final scenario, which goes by the flamboyant yet inscrutable title "vacuum metastability event." It is also referred to as the "collapse of the vacuum," but that's really not much of an improvement. After all, the process we're discussing here involves quantum mechanics, and that makes it less than intuitive from the outset, no matter what you call it. In fact, I considered omitting it, primarily because, rather than being a cosmic connection, it is more of a cosmic *disconnection*, a one-off event, to be sure. But in keeping with the premise that the universe can move in mysterious ways heretofore unforeseen, I include it here.

In simpler times, the universe was considered to be expanding, and not only expanding but slowing down as well due to the gravitational drag of its matter and dark-matter content. In an expanding but decelerating universe, galaxies would be conveyed from other galaxies at an increasingly slower rate. Therefore astronomers assumed that when telescopes could finally measure exploding stars called Type Ia supernovae from cosmological distances (several billion light-years or so), they would exhibit a predictable decrease in peak brightness with increasing distance. Little did they know that the big bang had an adjunct force that would have to be reckoned with. By 1998, when they had stitched together enough observations of distant supernovae to measure the expansion rate, they found that supernovae were 10 to 20 percent *fainter* than predicted in a gently decelerating universe, meaning that stars and galaxies were being borne away at an increasing, not decreasing, rate. Given the look-back times of the supernovae, they concluded that the universe has been accelerating for the past 5 billion years.

Overnight, that discovery changed the face of cosmology as it was suddenly realized that Einstein had been right way back in 1915. That's when he inserted a mathematical term in his field equations that endowed the vacuum of space with energy that acted to stabilize his model of a closed, spherical universe, by keeping it from gravitationally collapsing on itself. He called this antigravity term the cosmological constant, even though he himself had no explanation for it. Of course, many are now familiar with the lore about how Einstein withdrew the term after it was shown that the universe was expanding and thus had no use for a negative pressure, calling it "theoretically unsatisfactory."* Astronomers today, when speaking in context of an accelerating universe, call this vacuum energy "dark energy," and it is one of the biggest puzzles facing modern cosmological research.

So the vacuum of space, it turns out, is not so vacuous after all but is, as physicists have come to know it, a highly structured medium that, like frozen or vaporous states of water, can theoretically come in a variety of different energy levels. Quantum theory holds that the vacuum in our universe teems with elementary particles that have the ability to constantly pop into and out of existence like bubbles in seltzer water. This particle-popping process imparts the energy that constantly drives acceleration on top of expansion, or so goes the theory. Thus, our vacuum is said to have a "positive energy" (albeit very small). This small, positive energy state means one of two things for our vacuum. Precisely because of its low energy, most physicists refer to it as the "true" vacuum because it is unlikely to decay to an even lower, less energetic state. However, for the same reason, other physicists speculate that its positive energy, though small, nonetheless qualifies it as a "false vacuum," one that is unstable (or metastable) and hence capable of decaying to a lower energy level should it be disturbed in some way.

What kind of "something" could so disturb the vacuum that it would trigger a metastable event? Imaginative scientists and science fiction writers have proposed a number of possibilities. One thought that attracted a lot of media attention is that a new generation of elementary particle accelerators might one day initiate

---

*Einstein never called the cosmological constant his "biggest blunder," as is often stated.

a high-energy "heavy" ion event that would penetrate the "barrier" separating the false and true vacua, releasing a fireball of elementary particles that would expand throughout the universe at close to the speed of light.[21] It would mean nothing less than the deletion of the universe, or in the words of particle physicists Sidney Coleman and Frank De Luccia in a 1980 paper: "Vacuum decay is the ultimate ecological catastrophe; in the new vacuum there are new constants of nature; after vacuum decay, not only is life as we know it impossible, so is chemistry as we know it."[22]

So now you see why I call this a cosmic disconnection; the universe unplugged. But the question remains, could such an event happen, or, more to the point, could the universe do this spontaneously? In both cases, say most physicists, the answer (assuming their theories are correct) is no. The vacuum today is the one true vacuum, despite its having positive energy. Still, it's worth noting that, according to the standard theory of inflation, the universe has done this kind of thing before, just a fraction of a second after the big bang when the universe expanded exponentially by an extraordinary factor, as much as 10^{50} times. In that nano-wink of an eye after the big bang and before the onset of inflation, the universe was in a false vacuum energy state made up of a field of particles with energy density values that varied throughout space. In some regions, the energy density was high and in others it was low, with the lowest regions corresponding to the true vacuum. Such a particle landscape would resemble a rolling countryside of peaks and valleys. Our universe arose when it was nudged from one of the high-density regions and rolled "downhill" to a lower energy state, inflating exponentially as it did so. In the appended theory of inflation, called "eternal" inflation, this process goes on forever. (As a bonus, this theory also predicts that our universe may be one of an infinite number of separate, or "pocket," universes residing within a vast multiverse, also known as the megaverse, metaverse, omnium, and holocosom.) Whether our universe is the only one or one of many, should its energy state roll down another energy-density hill, or should it "tunnel" through the hill to reach a lower density state (say, via a particle-accelerator experiment), it would be all over for existence as we know it.

Don't hold your breath for this one. If such an event were going to happen, say particle physicists, it should have happened a long time ago, in fact, long before the invention of particle accelerators and particle physicists. Particle annihilation events equivalent to particle accelerator experiments have been going on since the formation of Earth. For example, high-energy cosmic rays have struck the surface of the Moon and other airless cosmic bodies for billions of years, as well as collided with other energetic particles in interstellar space without incident (otherwise we wouldn't be here). A report published in 2000 on behalf of the Brookhaven National Laboratory on Long Island, which at the time was constructing the Relativistic Heavy Ion Collider (RHIC)—which some theorists mused could trigger a new vacuum state via heavy ion collision experiments—concluded that possible particle-collision disasters would be triggered naturally at a rate that is at the very least 1,000 times higher than for high-energy particle accelerators. "It is clear," the authors of the report wrote, "that cosmic rays have been carrying out RHIC-like 'experiments' throughout the universe since time out of mind."[23]

RHIC has been in operation since 2000, and the universe's bubble still hasn't burst. Therefore, a catastrophe on par with a vacuum metastability event seems unlikely in the near term or, for that matter, for the next tens of billions of years.[24] Astrophysicists put the risk of such an event being triggered by particle accelerators over the next billion years at 1 in a trillion, which is about as unlikely as unlikely can be.[25] The quantum details supporting this claim are complex in the extreme, but the upshot appears to be that a collapse of the vacuum won't happen, not only because arguments to the contrary are often contrived and stretch the laws of physics beyond acceptable bounds, but also because, to our knowledge, one has yet to occur. Put in anthropic terms, we could say that the most probable region of space for us to find ourselves in is a region with an extremely small (but not too small) vacuum energy—or we wouldn't be here to report otherwise.

Scientists will continue to poke and probe dark energy, trying to determine its strength, or pressure, versus its density, or pervasiveness, throughout the universe—for that will tell them more about how acceleration and the vacuum may change in the far future. There are at least three possibilities so far. If more precise cosmological observations indicate that acceleration is a transient phenomenon and may trail off and finally go away altogether one day, gravity could reassert itself and collapse the universe within a time frame that is comparable to its present age. On the other hand, if it is the nature of dark energy to become stronger rather than weaker over time, it will continue to drive acceleration at ever-increasing rates. Eventually, in 20 billion years or so, the cosmos and everything in it could end in a so-called big rip, in which all bound systems—including galaxies, stars, planets, moons, people, and atoms—would literally be pulled apart.[26]

Between these two extremes is the "fade away" scenario: acceleration continues apace or flattens out, and the universe, very much dimmer and more diffuse, continues expanding with its vacuum intact for tens of billions of years. However you look at it, the universe appears destined to disconnect or, perhaps, recycle itself eventually, whether humankind is here to experience the finality or not. And who knows? Perhaps we live in a recycled universe, one that has invented and reinvented itself in untold incarnations and harbored a succession of myriad life-forms that evolved, flourished, and died off in times that predate the big bang. Perhaps we are not only made of star stuff but the stuff of past universes and of times before and times before that.

CHAPTER 9

THE ULTIMATE
COSMIC CONNECTION

Any complex historical outcome—intelligent life on Earth, for example—
represents a summation of improbabilities and becomes thereby absurdly
unlikely.
  —Stephen J. Gould, *The Flamingo's Smile:*
    Reflections in Natural History

My intention in writing this book was to present a general review on how life—Darwinian mechanisms aside—is connected to events that happen in the near and far universe. Of course, this is not a new idea by any means, as is plain by the weather-beaten "we-are-star-stuff" gnomic that stands prominently in the middle of the whole discussion over astronomy and life. My intent, however, was to draw readers' attention to the many subtle, and not so subtle, astronomical processes that have played a role in aiding and abetting existence as we know it, or in how we think about existence as we know it. Moreover, I wanted to include the consideration of other possible intelligent life-forms, because we may one day hear from them—something that would be one of the biggest cosmic connections of all time.

Throughout the book, we've explored some astronomy basics like orbital mechanics, solar physics, stellar evolution, and cosmology, in ways that I hope readers have found a little more refreshing, and perhaps more intriguing, than the usual cut-and-dried textbook approach. It's one thing to consider what happens when a star explodes; it's quite another to contemplate the short- and long-term effects that the explosion might have on any life-forms in the vicinity, incipient or otherwise. Though we have come to gratuitously regard the Sun's output as being constant, that our solar neighborhood is a rather plain vanilla region of the Galaxy, and that Earth is a too-small target for asteroids and comets, our cosmic connections indicate otherwise by pointing out the lasting effects that the rare exceptions have had in Earth's past and may well have in the future.

But any meaningful meditation on how astronomical events influence life on Earth must eventually culminate in a final, simple question, one that has no simple answer if indeed it has an answer at all: how are we to interpret the fact that our cosmic connections have been very good to us, in fact, encouraged life and even spared us from destruction? The evidence of our good fortune presented in the foregoing chapters points to one obvious inference, which I have done my best not to overstate but which is unavoidable: that is, had our cosmic situation been slightly different—if the Sun varied a bit more in output than it does, if the asteroid that exploded over Tunguska in 1908 had been kilometers, not meters, across, if Earth's orbit brought it too close or too far from the Sun—we might not be here today to marvel at the universe we occupy and to wonder how or why we are here at all. What seems like an accumulation of dumb luck is made all the more astonishing because of how improbable our luck has been. One can't help but be wowed by all the fortuitous coincidences that directly affect life on our planet:

- That our Sun is a single, stable star when most stars have companions.
- That Earth's axis is held in check by the Moon.
- That Earth's orbit is nearly circular.
- That Earth is where it is, in the sweet spot of the habitable zone of the Sun.
- That our Sun is in the sweet spot of the habitable zone of the Galaxy.
- That our Galaxy is undisturbed by supernovae or star formation and has been since humans arose on this planet.
- That the low-dense, dust-free conditions of the local interstellar medium in the last several million years have been favorable to the development of life on Earth.

We can add to this list a slew of other astonishing cosmic coincidences, such as the fundamental constants that physics claims determine how the entire universe is constructed and how its components function. These include the speed of light in a vacuum, the gravitational constant, the masses of various elementary particles, the strong nuclear force (which holds protons and neutrons together in an atomic nucleus), the Planck constant (used to relate the energy of a quantum of radiation to that radiation's frequency), and the fine-scale constant (which governs the strength of the electromagnetic force). Slight variations in any of these "regulators" would preclude our existence. For example, the gravitational constant allows for the formation of stars like our Sun. If it were a little weaker, stars would be less massive and not burn hydrogen; any higher and they would be more massive and live far shorter lives. A slightly weaker strong nuclear force would prevent hydrogen from burning and elements heavier than helium wouldn't exist; any stronger, and hydrogen nuclei would bind together as stable particle creatures called "diprotons" (in the normal universe these pairs are unstable and rapidly decay). A slight change in either direction would prevent carbon from forming in red giant stars at just the right amounts necessary for carbon-based life.

If you stand back and look at the whole cosmic picture, it almost seems as if the control knobs of the universe have been twiddled just for us, or at least for the ascent of intelligent life. Everything had to be as right as Goldilocks's porridge, from the electron-proton mass ratio right up to the expansion rate of the universe. Even the energy of the vacuum energy, discussed in the previous chapter, appears to be just right according to cosmologists' observations of distant supernovae—not too weak, not too strong. Any weaker, and the universe would have collapsed under its own weight by now; any stronger, and it would have dispersed into oblivion before galaxies could have even formed.*

The notion describing this life-accommodating, finely tuned universe is called the anthropic cosmological principle. In both its weak and strong versions, as well as those that fall in the middle, it essentially holds that our universe comes ready-made with properties that allow life in some form to develop: from basic carbon-based life right on up to intelligent observers. Life lies at the entire center of the mechanistic universe, and, in the strong version of the anthropic principle, the universe is attuned to intelligent life as much as intelligent life is attuned to it. Or, as physicist John A. Wheeler wrote in the foreword to John D. Barrow and Frank J. Tipler's landmark book *The Anthropic Cosmo-logical Principle*, "It is not only that man is adapted to the universe. The Universe is adapted to Man."

---

*Physicists' theoretical predictions, however, indicate that the cosmological constant should be 120 orders of magnitude greater than it is! This discrepancy is known as the cosmological constant problem. And, indeed, it is.

The anthropic principle was first coined by astrophysicist Brandon Carter in 1973 during a symposium held in Kraków, Poland, honoring, ironically enough, Copernicus on his five-hundredth birthday. Carter was reacting to what he saw as "exaggerated subservience to the 'Copernican principle,'" which holds that humans are in no way special to the universe. Carter was okay with the idea that humanity didn't occupy a privileged central place in the universe, but what he objected to was interpreting the Copernican principle to mean that our situation cannot be privileged in *any* sense. This "dogma," said Carter, was untenable if one accepts that favorable conditions in the universe are a prerequisite for our existence, that the universe evolves and, on local scales anyway, can vary in structure and content that are conducive, or not, to life. Carter stated his point this way: "[W]hat we can expect to observe must be restricted by the conditions necessary for our presence as observers."[1] In many respects he was stating the obvious—that a universe absent of living observers is unobservable—but his point struck a chord that still resonates with physicists and cosmologists today.

Without doubt, the existence of carbon-based life, and particularly intelligent carbon-based life, requires some very special preconditions, including among other things a habitable planet, a long-lived star, and physical constants that have specific values. But acknowledgment of these requirements leaves us with two sticky assumptions if we accept that our universe is the only one that exists: either we are very lucky for no apparent reason or we are lucky by design. Recently, a third assumption has been broached by physicists working on the idea that our universe is one of countless others in a vast multiverse: given a limitless number of universes, each with the constants of nature tuned randomly to other values, ours must naturally be the one "experiment" in which the constants are fine-tuned for life.[2] How many experiments would you need to explain the perfect universe we live in? String theorists, who champion this idea, say 10^{500} would suffice.[3]

Unfortunately, the anthropic principle is often used to prop up such nonscientific concepts as creationism or intelligent design. Theologians and theistic philosophers, too, are occasionally guilty of installing an anthropic brace here and there, or even an entire foundation, to their ideologies, particularly when they argue that a universe capable of producing intelligent life confirms the existence of God.[4] In its purest form, the anthropic principle does not default to a "supermind" for which there can be no proof. On the other hand, for some it does have its theistic nuances, particularly when advocates claim the universe is "self-aware" or when they cite the biblical scripture Psalm 8:4 as a reference: "What is man that Thou art mindful of him?"

Many scientists have gone on record stating they have no use at all for the anthropic principle, and not strictly because of its religious overtones. It isn't science, they argue. Why? Because it cannot make any falsifiable predictions.[5] In other words, there is no way to prove the anthropic principle wrong. It's more of a postscript that calls on the unknown (a supernatural grand designer, for example) to account for the known properties of the universe. It may be a teleological out, say critics, but it is not science.

For example, is it really necessary to invoke the anthropic principle to explain something like the isotropy of the universe or the value of the vacuum energy? It's true that a small set of initial conditions are responsible for why galaxies are evenly distributed across the universe rather than lumped in one region, and it's true that the vacuum energy is such that runaway inflation hasn't destroyed the universe, but why make the anthropic leap? Physicist Heinz R. Pagels, who called the principle "much ado about nothing," quipped that the bigger question was "how can such a sterile idea reproduce itself so prolifically?"[6]

More seriously, warn other scientists, invoking anthropic reasoning in theories may not only inhibit and even undermine the tried-and-true scientific method of formulating and testing hypotheses, but it could also end up dividing the scientific community (primarily theoretical physicists) into various camps of beliefs without consensus.[7]

Although I think it unlikely that the anthropic principle will subvert the foundations of science, it might weaken it in places, particularly the field of cosmology, in which many other scientific disciplines reflected in the principle's tenets, such as physics, quantum mechanics, chemistry, and biology, merge and move as one. As a humble science scribbler, I must admit that I sometimes feel conflicted over how much weight science sometimes seems to give the anthropic principle. It is neither science nor philosophy, but an amalgam of both. It is also an intriguing intellectual exercise that provides us with a unique perspective and certainly gives us something to think about by demonstrating how truly interconnected the universe is, particularly at subatomic scales. On the other hand—and this is where more strenuous objections often arise—it can be overly anthropocentric, especially when its proponents assume that all intelligent life must resemble, in broad terms, life on this planet.

When I consider the long history of our world and that of our solar and galactic environment in context with the brevity of the human epoch, I'll be the first to declare, "Hey, it's amazing we're here!"—because it *is* amazing we're here. But our amazement is a natural reaction to realizing just how outrageous our fortune has been. As paleontologist Stephen J. Gould put it, "[S]omething has to happen, even if any particular 'something' must stun us by its improbability."[8] I see no reason to draw anthropic conclusions from such improbabilities, unless one wants to argue for the existence of a supernatural designer (i.e., God), in which case we are no longer talking about science.

What is true? I think this is true: whatever the explanation may be for fine-tuning, when all is said and done, the only reason *we* are here instead of some other life-form, like dinosaurs, is because no cosmic mass-extinction event of significant magnitude has happened during the evolution of our species in our immediate region of the Milky Way for millions of years. That's not to say parts of the world haven't been affected by celestial events or that even evolution itself hasn't, perhaps, been tweaked now and again by a solar deviation here or a well-placed asteroid there; but it's obvious that none have been serious enough to prevent the ascent of humankind. We can give a nod to factors such as the nuclear force, the formation of carbon, the lifetimes of stars, and even the gravitational constant, but our success—or that of any other life-forms that may exist in this universe—is in no small way a product of happy coincidences.

Each year it seems, astronomers turn up further evidence of our improbable luck. A good recent example concerns the Moon, an object we give no more daily thought to than a radish, and yet its exact behavior, mass, and placement are responsible for making Earth such a stable planet. In 2007, astronomers used the Spitzer Space Telescope to look for dusty debris disks around 400 stars, each between 30 million and 50 million years old. This is about the same age that the Sun would have been when Earth, so goes the theory, was struck by a Mars-size object—the debris of which was thought to have formed the Moon. Such a massive collision would produce great quantities of dust around these stars, which Spitzer would detect as an excess in infrared brightness. The astronomers, however, found only 1 star out of 400 immersed in a dusty disk extensive enough to have resulted from a huge planet-planet collision, like the one that formed the Moon. If we consider these 400 stars a fair sample, then we can assume that satellites like the Moon are uncommon in the universe, appearing in 5 to 10 percent of planetary systems.[9] Of course, we have our Moon to thank for preventing our planet's axis from gyrating wildly from its almost up-down orientation, something that would have severely disrupted the Earth's climate and any developing life. In auspicious anthropic terms, we might say that it should come as "no surprise" that intelligent life would arise on a planet fortunate enough to have its rotational axis stabilized by a large natural satellite over a period of several billion years. At the same time, given the kind of cosmic calamities that can occur over the lifetime of a planet, how surprised would we be to find that, on average, intelligent life could just as easily fail to emerge on a planet with a stable axis?

By the same reasoning, we should not be surprised to discover that we live in an exceptionally quiet galaxy.[10] Compared to its neighbors, astronomers say, the Milky Way is rather humdrum, largely quiescent, and free of a too-traumatic past in which it might have suffered significant mergers. Even our nearest large neighbor, the Andromeda Galaxy, judging by the population of young stars found in its halo (typically a realm for older stars), has been influenced by mergers over the last 8 billion years. The Milky Way, however, has been largely undisturbed by such occurrences, thus providing the Sun with a serene setting for harboring planets and, of course, life.

Our luck continues to hold. A statistical analysis conducted a few years ago indicates our solar system may be the result of a simple twist of fate. In comparing the solar system with other known exoplanetary systems, it's obvious that most don't even come close to resembling the solar system. For one thing, the planets in our solar system have circular orbits, or nearly so, unlike those in most known exoplanetary systems. For another, the major planets of exoplanetary systems are often found close to the central star—in some cases too close for their presence to be explained by conventional planetary formation theories.[11] More recent discoveries have turned up planets that are a bit more comparable in mass to the Earth (if you consider five Earth masses comparable), but these worlds either orbit too close or too far from their host star to be considered promising candidates for life.

Why should our planetary system be so special? Is it because everything in the universe is finely adjusted so that the universe and the solar system comprise the just-right porridge for the habitation of life? Or could it be that the Sun formed in a special way—not in a cluster of stars, as most stars do, but as a single star? As such, it would never have suffered encounters with other stars or hooked up with a companion, which could have stirred up the orbits of the planets and even ejected them.[12]

In truth, astronomers have no idea whether the solar system is unique or not, but a recent finding may be telling. In the February 15, 2008, issue of the journal *Science*, an international team of researchers reported the discovery of a "scaled" version of the solar system, 4,800 light-years away in the direction of the constellation Scorpius. Although the host star itself has only half the mass of the Sun and is slightly cooler, it harbors two planets that resemble Jupiter and Saturn. The innermost planet has a mass about 71 percent that of Jupiter, while the outermost planet is some 90 percent the mass of Saturn. Moreover, the distances of the two planets, while not an exact match to those of Jupiter and Saturn, are nonetheless proportional to them. The team calculated that the inner planet lies 345 million kilometers (214 million miles) from its star while the outer planet's distance is 690 million kilometers (430 million miles). Compare this with Jupiter, which lies 780 million kilometers (485 million miles) from the Sun, and Saturn, which is 1.4 billion kilometers (870 million miles) from the Sun. The implication, of course, is that there is still plenty of room between the host star and the innermost planet for one or perhaps two earthlike bodies.[13]

As search techniques improve in sensitivity, a variety of exoplanetary systems are bound to be discovered and surveyed. Based on the above solar system analog, and if you go by the where-there's-one-there's-two assumption, it may turn out that solar systems like our own aren't that unique after all—except for the fact that we happen to live in this one. For that reason alone, ours may stand as a singular model for quite some time, but, perhaps, not always. Even science fiction teaches us that life could conceivably be found in many evolutionary stages on other worlds and in variously arranged solar systems without invoking navel-gazing anthropic mechanisms. The reason: life is manifestly opportunistic and can happen in spite of, as much as because of, biological conditions.

Life also has pretty good timing. A closer look at some of the unique conditions or coincidences to which we owe our existence shows that, although most seem life-friendly, in fact, they really aren't over the long haul. Carbon-based life can arise only when all the tumblers fall into place, but how long that life sticks around depends on how quickly and how much the tumblers change.*

For example, we say that the Earth's axis is held in check by the Moon, and if not for the Moon we wouldn't be here. But as is well

---

*This raises an intriguing question: why would an intelligent designer incorporate random variables into a universe that could create life on the one hand and whimsically destroy it on the other?

known, the Moon's orbit is gradually moving farther away from the Earth. In a billion years or so, its steadying hold on obliquity will begin to relax, causing the tilt to increase, eventually reaching extremes of as much as 60°. If the current small variations in the Earth's obliquity can generate entrenched ice ages, imagine how larger and larger variations stepping up to magnitudes such as this over the next 100 million years could play havoc with the climate.

We say the Earth's orbit is nearly circular, and if that weren't true, we wouldn't be here either. But as we saw at the beginning of this book, over periods of 100,000 and 400,000 years it goes through cycles during which it becomes more significantly out-of-round, thus changing the climate by altering the amount of solar energy that reaches the ground during perihelion. This means ice ages are destined—programmed!—to happen again.

We say the Sun occupies a part of the Milky Way that is undisturbed by supernovae, but the Sun is continuously moving through interstellar space, and eventually, perhaps as soon as a few tens of thousands of years from now—a period nearly comparable to how long humankind has been looking at the stars—the environment in which it finds itself may not be so hospitable to life on this planet.

We say we're fortunate in that major asteroid impacts that may cause mass extinctions are rare events on Earth. But their infrequency is canceled out by the unprecedented level of devastation a single event could cause. It need happen only once to severely cripple or put an end to civilization as we now know it.

Finally, some scientists theorize that even the fundamental physical constants of nature may turn out not to be so constant over the lifetime of the universe.[14] If that's true, then combinations of a variety of constants could result in different types of universes containing very different forms of life and consciousness. Clouds of particles could be self-aware, as could gobbets of antimatter. In any case, in anthropic terms, the fine-tuning settings for life in one universe could be very different in another, assuming there are other universes.

This universe we know exists, and we also know we occupy a special place in it—*now*—but, as I hope I've demonstrated throughout the pages of this book, that's not always been the case, and it's certainly not going to stay that way forever. The ultimate cutoff point occurs in another 5 billion years when our Sun dies and our planet is reduced to a cinder. Assuming civilization has a good long run and doesn't self-extinguish, however, we won't have to wait nearly that long for certain cosmic connections to intervene. Long before the Sun burns out, orbital forcing and protracted solar fluctuations will alter our climate many times over; asteroids large and small will have had their way with our planet; and the Sun will have wheeled around the Galaxy perhaps 15 or 20 times, drifting through all manner of interstellar environments along the way. Perhaps a maverick black hole or an unknown star will pass through the Oort Cloud, sending thousands of comets toward the inner solar system, some of which may fall upon the Earth.

But assuming our technology advances along with us and, more important, our wisdom as well, the time may come when we will no longer be vulnerable to whatever the cosmos throws at us. Rather than being passive by-products and bystanders in the universe, we may become its active participants, harnessing the energy of stars and planets, the minerals of asteroids, and the ices of comets to civilization's advantage. Cynics may scoff at such notions, but for those who desire to know more about the *whys* than the *hows* of our existence, the threads of an answer may be found among the stars. Humanity determining and taking charge of its own destiny in the universe would be the ultimate cosmic connection.

ACKNOWLEDGMENTS

No science writer is a failure who has good sources. The ones listed below are some of the best, and I am very much indebted to them for their help in writing this book.

Randy Cerveny, Clark Chapman, Geoff Chester, Kris Davidson, Brian Dennis, John Dubinski, Robert Duncan, Doug Erwin, Douglas Gies, Joseph Gurman, Mark Horvat, John Imbrie, Mikko Kaasalainen, Avi Loeb, Jean Meeus, Adrian Melott, Stefan Rahmstorf, Seth Redfield, Gerard Roe, Bradley Schaefer, Seth Shostak, Sami Solanki, Jill Tarter, and Bruce Tsurutani. I apologize if I've overlooked anyone.

I would like to further recognize William Blair for his Vela supernova remnant Web page (http://fuse.pha.jhu.edu/~wpb/hstvela/hstvela.html); Duane Dunkerson "the Space Guy" for his astute comments on the observational history of the Martian canals (http://www.thespaceguy.com/default.htm); John Gosling and his wonderful Web page dedicated to H. G. Wells's *The War of the Worlds* (http://www.war-ofthe-worlds.co.uk/index.html); and, of course, the NASA Astrophysics Data System (http://adswww

.harvard.edu/) and the astro-ph site maintained by Cornell University (http://arxiv.org/list/astro-ph/new).

For help with illustrations (even though some of them didn't make it into the book), I'd like to thank Anna Lawford, photograph curator of the Alpine Club Photo Library; Jerry and Sandy Grulkey, for supplying a cache of classic science fiction illustrations of Mars; Christine Ramseyer and Gudula Metze of the Kuntsmuseum Basel in Switzerland, for providing the painting of the Mer de Glace by Samuel Birmann; and illustrator Nik Spencer, for his fine diagrams of the Milankovitch cycles in chapter 1.

Finally, a big thank you to my long-suffering wife, Alexandra Witze, who read the manuscript (several times) and provided her usual incisive and invaluable comments. I don't know how I can ever repay her, but I am certain she will find a way.

NOTES

INTRODUCTION

1. Douglas M. Hudgins, Charles W. Bauschlicher Jr., and L. J. Allamandola, "Variations in the Peak Position of the 6.2 mm Interstellar Emission Feature: A Tracer of N in the Interstellar Polycyclic Aromatic Hydrocarbon Population," *Astrophysical Journal* 632 (2005): 331.

2. M. J. Mumma et al., "Organic Composition of C/1999 S4 (LINEAR): A Comet Formed Near Jupiter?" *Science* 292 (2001): 1334.

CHAPTER 1. TILT-A-WHIRL WORLD

1. John McPhee, *Annals of the Former World, Book 2: In Suspect Terrain* (New York: Farrar, Straus and Giroux, 1998), p. 254.

2. Louis Jean Agassiz, "Evidence of a Glacial Epoch," in *A Source Book in Geology, 1400–1900*, ed. Kirtley F. Mather and Shirley L. Mason (Cambridge, MA: Harvard University Press, 1967), p. 329.

3. McPhee, *Annals of the Former World*, p. 257.

4. James Rodger Fleming, "James Croll in Context: The Encounter between Climate Dynamics and Geology in the Second

Half of the Nineteenth Century," *Milutin Milankovic Anniversary Symposium: Paleoclimate and the Earth Climate System*, August 30–September 2, 2004 (2005): 46, http://www.meteohistory.org/2006 historyofmeteorology3/3fleming_croll.pdf (accessed February 6, 2008).

5. Ibid., p. 47.

6. Vasko Milanković, *Milutin Milanković: 1879–1958* (Katlenburg-Lindau, Germany: European Geophysical Society, 1995), p. 131.

7. Ibid., p. 132.

8. Ibid., p. 5.

9. Milankovitch's colleague, the German climatologist Vladimir Koppen, was Wegener's father-in-law.

10. Patrick Hughes, "The Meteorologist Who Started a Revolution," http://www.pangaea.org/wegener.htm (accessed February 7, 2008).

11. Jean Meeus, *Astronomical Algorithms* (Richmond, VA: Willmann-Bell Inc., 2005), p. 131.

12. Ibid., p. 131.

13. Milanković, *Milutin Milanković*, p. 52.

14. J. W. Head et al., "Tropical to Mid-latitude Snow Ice Accumulation, Flow and Glaciation on Mars," *Nature* 434 (2005): 347.

15. Gerard Roe, "In Defense of Milankovitch," *Geophysical Research Letters* 33 (2006): L24703.

16. Gerard Roe, e-mail to author, May 25, 2007.

CHAPTER 2. AN IMPERFECT SUN

1. Unless otherwise noted, the quotes in this section are derived from Emmanuel Le Roy Ladurie, "The Problems of 'the Little Ice Age,'" chap. 4, in *Times of Feast, Times of Famine: A History of Climate since the Year 1000*, trans. Barbara Bray (Garden City, NY: Doubleday, 1971), p. 146.

2. Ibid., p. 170.

3. S. U. Nussbaumer, H. J. Zumbühl, and D. Steiner, "Fluctuations of the Mer de Glace (Mont Blanc Area, France) AD 1500–2050: An Interdisciplinary Approach Using New Historical Data and Neural Network Simulations," *Zeitschrift für Gletscherkunde und Glazialgeologie* 40 (2007): 39–42.

4. Ladurie, *Times of Feast, Times of Famine*, p. 197.

5. F. E. Matthes, "Report of the Committee on Glaciers," *Transactions of the American Geophysical Union* 20 (1939): 518.

6. Brian Fagan, *The Little Ice Age: How Climate Made History* (New York: Basic Books, 2000), p. 91.

7. J. Oerlemans, "Extracting a Climate Signal from 169 Glacier Records," *Science* 308 (2005): 675.

8. Lloyd Burckle and Henri D. Grissino-Mayer, "Stradivari, Violins, Tree Rings, and the Maunder Minimum: A Hypothesis," *Dendrochronologia* 21 (2003): 41.

9. John A. Eddy, "The Maunder Minimum," *Science* 192 (1976): 1189.

10. Joann D. Haigh, "The Effects of Solar Variability on the Earth's Climate," *Philosophical Transactions of the Royal Society* 361 (2003): 95.

11. N. Scafetta and B. J. West, "Estimated Solar Contribution to the Global Surface Warming Using the ACRIM TSI Satellite Composite," *Geophysical Research Letters* 32 (2005): L18713.

12. S. K. Solanki et al., "Unusual Activity of the Sun during Recent Decades Compared to the Previous 11,000 Years," *Nature* 431 (2004): 1084.

13. Jorge Meléndez, Katie Dodds-Eden, and José A. Robles, "HD 98618: A Star Closely Resembling Our Sun," *Astrophysical Journal* 641 (2006): L133.

CHAPTER 3. A TEMPERAMENTAL SUN

1. P. R. Barnes et al., "Electric Utility Industry Experience with Geomagnetic Disturbances," Oak Ridge National Laboratory, ORNL-6665 (November 25, 1991): 61–66.

2. "Geomagnetic Storms: Reducing the Threat to Critical Infrastructure in Canada," Threat Analysis by the Office of Critical Infrastructure and Emergency Preparedness (April 25, 2002): TA02–001.

3. J. A. Eddy, "A Nineteenth-Century Coronal Transient," *Astronomy & Astrophysics* 34 (1974): 235. (Note: The 1860 eclipse and Tempel's observations were described by C. A. Raynard in *Memoirs of the Royal Astronomical Society* 41 [1879]: 520.)

4. Brian Dennis, "Which Are More Powerful, Flares or CMEs?" http://sprg.ssl.berkeley.edu/~tohban/nuggets/?page=article&article_id=10 (accessed May 2007).

5. Graham R. Thompson and Jonathan Turk, *Earth Science and the Environment* (Belmont, CA: Brooks/Cole–Thomson Learning, Inc., 2005), p. 52.

6. SOHO LASCO CME Catalog, http://cdaw.gsfc.nasa.gov/CME_list/UNIVERSAL/2003_10/univ2003_10.html (accessed July 2007).

7. S. Yashiro et al., "Spatial Relationship between Solar Flares and Coronal Mass Ejections," *Astrophysical Journal* 673 (2008): 1174.

8. V. N. Ishkov, "Evolution and Flare Productivity of Active Regions in October–November 2003," *Solar System Research* 40 (2006): 117.

9. NOAA, "Space Environment Center Preliminary Report and Forecast of Solar Geophysical Data," *Rep. 1471* (2003), http://sec.noaa.gov/weekly.html (accessed July 2007).

10. David Brodrick, Steven Tingay, and Mark Wieringa, "X-ray Magnitude of the 4 November 2003 Solar Flare Inferred from the Ionospheric Attenuation of the Galactic Radio Background," *Journal of Geophysical Research* 110 (2005): A09S36.

11. R. C. Carrington, "Description of a Singular Appearance Seen in the Sun on September 1, 1859," *Monthly Notices of the Royal Astronomical Society* 20 (1859): 13.

12. B. T. Tsurutani et al., "The Extreme Magnetic Storm of 1–2 September 1859," *Journal of Geophysical Research* 108 (2003): 1268.

13. B. C. Thomas, C. H. Jackman, and A. L. Melott, "Modeling Atmospheric Effects of the September 1859 Solar Flare," *Geophysical Research Letters* 34 (2007): L06810.

14. Tsurutani et al., "The Extreme Magnetic Storm of 1–2 September 1859," p. 1268.

15. Bruce Tsurutani, e-mail message to author, March 8, 2006.

16. Rachel A. Osten et al., "Nonthermal Hard X-ray Emission and Iron Ka Emission from a Superflare on II Pegasi," *Astrophysical Journal* 654 (2007): 1052.

17. Eric P. Rubenstein and Bradley E. Schaefer, "Are Superflares on Solar Analogues Caused by Extrasolar Planets?" *Astrophysical Journal* 529 (2000): 1031.

18. Jeffrey C. Hall, Gregory W. Henry, and G. Wesley Lockwood, "The Sun-Like Activity of the Solar Twin 18 Scorpii," *Astronomical Journal* 133 (2007): 2206.

19. E-mail from Department of Homeland Security Office of Multimedia to the author, June 20, 2007.

20. "National Critical Infrastructure Protection Research and Development Plan," Department of Homeland Security, Science and Technology Directorate (2004): 53.

21. Hugh S. Hudson, "The Unpredictability of the Most Energetic Solar Events," *Astrophysical Journal* 663 (2007): L45.

22. Ken Caldeira and James F. Kasting, "The Life Span of the Biosphere Revisited," *Nature* 360 (1992): 721.

23. Timothy M. Lenton and Werner von Bloh, "Biotic Feedback Extends the Life Span of the Biosphere," *Geophysical Research Letters* 28 (2001): 1718.

CHAPTER 4. AT ANY TIME

1. For a selection of videos of this event, see http://aquarid .physics.uwo.ca/~pbrown/Videos/peekskill.htm (accessed February 19, 2008).

2. Thomas Graf et al., "Size and Exposure History of the Peek-skill Meteoroid," *Meteoritics* 29 (1993): 469.

3. P. Brown, "The Orbit and Atmospheric Trajectory of the Peek-skill Meteorite from Video Records," *Nature* 367 (1994): 624.

4. Kevin Yau, Paul Weissman, and Donald Yeomans, "Meteorite Falls in China and Some Related Human Casualty Events," *Meteoritics* 29 (1994): 864.

5. William A. Cassidy, "Estimated Frequency of a Meteorite Striking an Aircraft," NTSB Final Report (December 17, 1997): 1, http://www.dartmouth.edu/~chance/teaching_aids/books_articles/C assidy.pdf (accessed June 2007).

6. I. P. Wright, M. Grady, and A. Sexton, "The Chemistry of Impacting Bodies Recorded on Eureca," *Meteoritics* 30 (1995): 602.

7. D. Lear et al., "STS-118 Radiator Impact Damage," *Orbital Debris Quarterly News* 12 (2008): 3.

8. "Orbital Box Score," *Orbital Debris Quarterly News* 12 (October 2008): 12. Updates available online at http://orbitaldebris.jsc.nasa .gov/newsletter/newsletter.html (accessed October 22, 2008).

9. "Cosmos 954: An Ugly Death," *Time*, February 6, 1978, http://www.time.com/time/magazine/article/0,9171,945940-1,00 .html (accessed February 13, 2008).

10. A fascinating firsthand account of the recovery operation of Cosmos 954 by Quentin Bristow, who was with the Geological Survey of Canada at the time, can be found at http://gsc.nrcan.gc.ca/ gamma/ml_e.php#bg (accessed February 13, 2008).

11. "Fengyun-1C Debris: One Year Later," *Orbital Debris Quarterly News* 12 (2008): 2.

12. Ian Halliday, Alan T. Blackwell, and Arthur A. Griffin, "The Frequency of Meteorite Falls on the Earth," *Science* 223 (1984): 1405.

13. I. Halliday, A. T. Blackwell, and A. A. Griffin, "Meteorite Impacts on Humans and Buildings," *Nature* 318 (1985): 317.

14. Steven R. Chesley, "Potential Impact Detection for Near-Earth Asteroids: The Case of 99942 Apophis (2004 MN4)," in *Proceedings of the 229th Symposium of the International Astronomical Union* 1 (2005): 215.

15. J. D. Giorgini et al., "Predicting the Earth Encounters of (99942) Apophis," *Icarus* 193 (2008): 1–19.

16. "Dealing with the Threat of an Asteroid Striking the Earth," position paper, American Institute of Aeronautics and Astronautics, April 1990, http://pdf.aiaa.org/downloads/publicpolicyposition papers/Asteroid-1990.pdf (accessed July 2007).

17. David Morrison, "Tunguska Impactor Size Revision," http://impact.arc.nasa.gov/news_detail.cfm?ID=179 (accessed February 8, 2008).

18. Roy A. Gallant, "Journey to Tunguska, *Sky & Telescope* 87 (1994): 38.

19. Luigi Foschini, "A Solution for the Tunguska Event," *Astronomy & Astrophysics* 342 (1999): L1.

20. V. A. Bronshten, "Nature and Destruction of the Tunguska Body," *Planetary and Space Science* 48 (2000): 855.

21. P. Farinella et al., "Probable Asteroidal Origin of the Tunguska Cosmic Body," *Astronomy & Astrophysics* 377 (2001): 1081.

22. L. Gasperini et al., "A Possible Impact Crater for the 1908 Tunguska Event," *Terra Nova* 19, no. 4 (2007): 245–51.

23. L. W. Alvarez et al., "Extraterrestrial Cause for the Cretaceous Tertiary Extinction," *Science* 208 (1980): 1095.

24. Donald K. Yeomans, "The Tunguska Event and the History of Near-Earth Objects," American Astronomical Society Meeting 209 (2006): 108.01.

25. Clark R. Chapman and David Morrison, "Impacts on the Earth by Asteroids and Comets: Assessing the Hazard," *Nature* 367 (1994): 33.

26. Clark Chapman, "Overview of Asteroid Impact Phenomenology," presentation given at the Planetary Defense Conference, March 7, 2007, Washington, DC. Available online at http://www.aero

.org/conferences/planetarydefense/2007papers/S4-1—Chapman-Brief.pdf (accessed February 8, 2008).

27. H. R. 1022, "To Provide for a Near-Earth Object Survey Program to Detect, Track, Catalogue, and Characterize Certain Near-Earth Asteroids and Comets," 109th Congress, 1st Session (2005).

28. "Summary and Recommendations from the 2007 Planetary Defense Conference," http://www.planetarydefense.info/resources/pdf/conference_white_paper.pdf (accessed August 18, 2007).

29. Edward T. Lu and Stanley G. Love, "Gravitational Tractor for Towing Asteroids," *Nature* 438 (2005): 177.

30. Steven R. Chesley et al., "Direct Detection of the Yarkovsky Effect by Radar Ranging to Asteroid 6489 Golevka," *Science* 302 (2003): 1739.

31. Joseph N. Spitale, "Asteroid Hazard Mitigation Using the Yarkovsky Effect," *Science* 296 (2002): 77.

32. D. I. Steel et al., "Are Impacts Correlated in Time?" *Hazards Due to Comets and Asteroids*, ed. T. Gehrels (Tucson: University of Arizona Press, 1994), p. 463.

33. John T. Wasson, "Large Aerial Bursts: An Important Class of Terrestrial Accretionary Events," *Astrobiology* 3 (2003): 163.

34. The Meteoritical Society's Meteoritical Bulletin Database, http://tin.er.usgs.gov/meteor/metbull.php?code=45817 (accessed February 19, 2008).

35. Spaceweather.com, October 8, 2007, http://www.spaceweather.com/archive.php?month=10&day=08&year=2007&view=view (accessed February 8, 2008).

36. William F. Bottke, David Vokrouhlicky, and David Nesvorny, "An Asteroid Breakup 160 Myr Ago as the Probable Source of the K/T Impactor," *Nature* 449 (2007): 48.

37. R. B. Firestone et al., "Evidence for an Extraterrestrial Impact 12,900 Years Ago That Contributed to the Megafaunal Extinctions and the Younger Dryas Cooling," *PNAS* 104 (2007): 16016.

38. *The Effects of Nuclear War*, NTIS order PB-296946, Office of

Technology and Assessment (May 1979). Available at http://www.fas.org/nuke/intro/nuke/7906/ (accessed July 11, 2007).

39. James A. Marusek, "Comet and Asteroid Threat Impact Analysis," 2007 Planetary Defense Conference, http://www.aero.org/conferences/planetarydefense/2007papers.html (accessed August 17, 2007).

40. "The Tsar Bomba ('King of Bombs')," http://www.nuclearweaponarchive.org/Russia/TsarBomba.html (accessed August 16, 2007).

CHAPTER 5. KEEP YOUR DISTANCE

1. J. M. Stil and J. A. Irwin, "GSH 138-01-94: An Old Supernova Remnant in the Far Outer Galaxy," *Astrophysical Journal* 563 (2001): 819.

2. B. Cameron Reed, "New Estimates of the Solar-Neighborhood Massive Star Birthrate and the Galactic Supernova Rate," *Astronomical Journal* 130 (2005): 1652.

3. John Ellis and David N. Schramm, "Could a Nearby Supernova Explosion Have Caused a Mass Extinction?" *Proceedings of the National Academy of Sciences* 92 (1995): 235.

4. L.-Y. Zhu et al., "Deep, Low Mass Ratio Overcontact Binaries. II. IK Pegasi," *Astronomical Journal* 129 (2005): 2806.

5. Eugenie Samuel, "Supernova Poised to Go off Near Earth," New Scientist.com, May 23, 2002, http://www.newscientis.com/article.ns?id=dn2311 (accessed September 11, 2007).

6. G. J. Mathews, G. Herczeg, and D. S. P. Dearborn, "Evolutionary Tracks for Betelgeuse," *American Astronomical Society Meeting* 192 (1998): 67.03.

7. Floor van Leeuwen, "The Hipparcos Mission," *Space Science Reviews* 81 (2004): 201.

8. F. Elias et al., "OB Stars in the Solar Neighborhood II: Kinematics," *Astronomical Journal* 132 (2006): 1052.

9. Douglas Erwin, e-mail to author, May 21, 2006.

10. K. Knie et al., "^{60}Fe Anomaly in a Deep-Sea Manganese Crust and Implications for a Nearby Supernova Source," *Physical Review Letters* 93 (2004): 171103.

11. Narciso Benítez, Jesús Máiz-Apellániz, and Matilde Canelles, "Evidence for Nearby Supernova Explosions," *Physical Review Letters* 88 (2002): 081101.

12. Seth Redfield, e-mail to author, May 16, 2006.

13. A. D. Thackeray, "Nebulosity Surrounding Eta Carinae," *Observatory* 69 (1949): 31.

14. Kris Davidson et al., "An Unusual Brightening of Eta Carinae," *Astronomical Journal* 118 (1999): 1777.

15. Kris Davidson, e-mail to author, May 30, 2006.

16. K. Davidson and R. M. Humphreys, "Eta Carinae and Its Environment," *Annual Reviews of Astronomy and Astrophysics* 35 (1997): 1.

17. Nathan Smith, "The Structure of the Homunculus. I. Shape and Latitude Dependence from H2 and [FeII] Velocity Maps of Eta Carinae," *Astrophysical Journal* 644 (2006): 1151.

18. Kris Davidson, e-mail to author, May 30, 2006.

19. Neil Gehrels, Luigi Piro, and Peter J. T. Leonard, "The Biggest Explosions in the Universe," *Scientific American* 287 (2002): 84.

20. A. L. Melott et al., "Did a Gamma-Ray Burst Initiate the Late Ordovician Mass Extinction?" *International Journal of Astrobiology* 3 (2004): 55.

21. Andrew Fruchter quoted in press release, "Hubble Finds That Earth Is Safe from One Class of Gamma-Ray Burst," Space Telescope Science Institute, May 10, 2006, STScI-2006-20.

22. A. S. Fruchter et al., "Long Gamma-Ray Bursts and Core-Collapse Supernovae Have Different Environments," *Nature* 441 (2006): 463.

23. N. Gehrels, "A Short Gamma-Ray Burst Apparently Associated with an Elliptical Galaxy at Redshift $z = 0.225$," *Nature* 437 (2005): 851.

24. Govert Schilling, "Black Holes by the Million Litter the Galaxy," *New Scientist* 1829 (1992): 16.

25. R. E. Rutledge, D. B. Fox, and A. H. Shevchuk, "Discovery of an Isolated Compact Object at High Galactic Latitude," *Astrophysical Journal* 672 (2008): 1137.

26. SGR/AXP Online Catalog, http://www.physics.mcgill.ca/~pulsar/magnetar/main.html (accessed October 22 2008).

27. E. P. Mazets, S. V. Golenetskii, Y. A. Guryan, and V. N. Ilyinskii, "The March 1979 Event and the Distinct Class of Short Gamma Bursts: Are They of the Same Origin?" *Astrophysics and Space Science* 84 (1982): 173.

28. U. S. Inan et al., "Ionization of the Lower Ionosphere by Gamma-Rays from a Magnetar: Detection of a Low Energy (3–10 ke V) Component," *Geophysical Research Letters* 26 (1999): 3357.

29. Frederick J. Vrba et al., "The Discovery of an Embedded Cluster of High-Mass Stars Near SGR 1900 +14," *Astrophysical Journal* 533 (2000): L17.

30. K. Hurley et al., "An Exceptionally Bright Flare from SGR 1806-20 and the Origins of Short-Duration Gamma-Ray Bursts," *Nature* 434 (2005): 1098.

31. U. S. Inan et al., "Massive Disturbance of the Daytime Lower Ionosphere by the Giant Gamma-Ray Flare from Magnetar SGR 1806-20," *Geophysical Research Letters* 34 (2007): L08103.

32. Chryssa Kouveliotou, Robert Duncan, and Christopher Thompson, "Magnetars," *Scientific American* (February 2003): 34.

33. C. Kouveliotou, "An X-Ray Pulsar with a Superstrong Magnetic Field in the Soft Gamma-Ray Repeater SGR 1806-20," *Nature* 393 (1998): 235. A. Tiengo et al., "The Calm after the Storm: XMM-Newton Observation of SGR 1806-20 Two Months after the Giant Flare of 2004 December 21," *Astronomy & Astrophysics* 440 (2005): L63.

34. Robert Duncan, e-mail to author, May 6, 2006.

35. Brian C. Thomas, "Gamma-Ray Bursts and the Earth: Exploration of Atmospheric, Biological, Climatic, and Biogeochemical Effects," *Astrophysical Journal* 634 (2005): 509.

36. C. P. Burgess and K. Zuber, "Footprints of the Newly Discov-

ered Vela Supernova in Antarctic Ice Cores?" *Astroparticle Physics* 14 (2000): 1.

37. Nathan Smith et al., "SN 2006gy: Discovery of the Most Luminous Supernova Ever Recorded, Powered by the Death of an Extremely Massive Star Like Eta Carinae," *Astrophysical Journal* 666 (2007): 1116.

38. Nathan Smith, as quoted in "Star Dies in Monstrous Explosion," BBC News, May 8, 2007, http://news.bbc.co.uk/2/hi/science/nature/6633609.stm (accessed September 2007).

CHAPTER 6. THERE GOES THE NEIGHBORHOOD

1. E. T. Byram et al., "Rocket Observations of Extraterrestrial Far-ultraviolet Radiation," *Astronomical Journal* 62R (1957): 9.

2. J. Hartmann, "Investigations on the Spectrum and Orbit of Delta Orionis," *Astrophysical Journal* 19 (1904): 268.

3. Mary Lea Heger, "Stationary Sodium Lines in Spectroscopic Binaries," *Publications of the Astronomical Society of the Pacific* 31 (1919): 304.

4. E. E. Barnard, "On the Dark Markings of the Sky, with a Catalogue of 182 Such Objects," *Astrophysical Journal* 49 (1919): 1.

5. Max Wolf, "On the Dark Nebula NGC 6960," trans. Brian Doyle and Owen Gingerich, *Astronomische Nachrichten* 219 (1923): 109.

6. Hendrik C. van de Hulst, "The Solid Particles of Interstellar Space," *Recherches Astronomiques de l'Observatoire d'Utrecht* 11 (1949): 1.

7. Harold I. Ewen and Edward M. Purcell, "Radiation from Galactic Hydrogen at 1,420 MHz," *Nature* 168 (1951): 356.

8. Frank Bash, "The Present, Past, and Future Velocity of Nearby Stars: The Path of the Sun in 10^8 Years," in *The Galaxy and the Solar System*, ed. Roman Smoluchowski, John N. Bahcall, and Mildred S. Matthews (Tucson: University of Arizona Press, 1986), p. 42.

9. Priscilla Frisch, "The Galactic Environment of the Sun," *American Scientist* 88 (January–February 2000): 52.

10. Ibid.

11. Dimitri M. Mihalas and James J. Binney, *Galactic Astronomy—Structure and Kinematics* (San Francisco: Freeman and Company, 1981), p. 625.

12. Priscilla C. Frisch and Jonathan D. Slavin, "Short-Term Variations in the Galactic Environment of the Sun," in *Solar Journey: The Significance of Our Galactic Environment for the Heliosphere and Earth*, ed. Priscilla C. Frisch (Dordrecht: Springer, 2006), p. 133.

13. Excellent images and information about the Vela Supernova Remnant can be found at Bill Blair's Vela Supernova Remnant Page, http://fuse.pha.jhu.edu/~wpb/hstvela/hstvela.html (accessed September 2007).

14. P. C. Frisch and D. G. York, "Interstellar Clouds near the Sun," in *The Galaxy and the Solar System*, p. 100.

15. Frisch and Slavin, *Solar Journey*, p. 135.

16. T. W. Berghöfer and D. Breitschwerdt, "The Origin of the Young Stellar Population in the Solar Neighborhood—A Link to the Formation of the Local Bubble?" *Astronomy & Astrophysics* 390 (2002): 299.

17. Frisch and York, *The Galaxy and the Solar System*, p. 84.

18. B. E. Penprase, J. D. Rhodes, and E. L. Harris, "Optical Observations of the Draco Molecular Cloud," *Astronomy and Astrophysics* 364 (2000): 712.

19. Sten F. Odenwald and Lee J. Rickard, "Hydrodynamical Processes in the Draco Molecular Cloud," *Astrophysical Journal* 318 (1987): 702.

20. Frisch and Slavin, *Solar Journey*, p. 139.

21. Hans-Reinhard Müller et al., "Heliospheric Response to Different Possible Interstellar Environments," *Astrophysical Journal* 647 (2006): 1493.

22. Priscilla Frisch, "The Galactic Environment of the Sun," *American Scientist* 88 (2000): 52.

23. R. C. Reedy, J. R. Arnold, and D. Lal, "Cosmic-Ray Record in the Solar System," *Science* 219 (1983): 127.

24. K.S. Carslaw, R.G. Harrison, and J. Kirkby, "Cosmic Rays, Clouds, and Climate," *Science* 298 (2002): 1732.

25. Nigel D. Marsh and Henrik Svensmark, "Low Cloud Properties Influenced by Cosmic Rays," *Physical Review Letters* 85 (2000): 5004.

26. D.R. Gies and J.W. Helsel, "Ice Age Epochs and the Sun's Path through the Galaxy," *Astrophysical Journal* 626 (2005): 844.

27. Nir J. Shaviv and Ján Veizer, "Celestial Driver of Phanerozoic Climate?" *GSA Today* 13 (2003): 4.

28. Knud Jahnke, "On the Periodic Clustering of Cosmic Ray Exposure Ages of Iron Meteorites" (2005): arXiv0504155J.

29. Harlow Shapley, "Note on a Possible Factor in Changes of Geological Climate," *Journal of Geology* 29 (1921): 502.

30. "The Milky Way Shaped Life on Earth," Danish National Space Center press release, December 2006, http://spacecenter.dk/research/sun-climate/other/the-milky-way-shaped-life-on-earth (accessed September 2007).

31. Mike Lockwood and Claus Fröhlich, "Recent Oppositely Directed Trends in Solar Climate Forcings and the Global Mean Surface Air Temperature," *Proceedings of the Royal Society A* 463 (2007): 2447.

32. Gies and Helsel, "Ice Age Epochs," p. 846.

33. Erik M. Leitch and Gautam Vasisht, "Mass Extinctions and the Sun's Encounters with Spiral Arms," *New Astronomy* 3 (1998): 51.

34. Alan Cutler, *The Seashell on the Mountaintop* (New York: Dutton, 2003), p. 138.

35. David M. Raup and J. John Sepkoski Jr., "Periodicity of Extinctions in the Geologic Past," *Proceedings of the National Academy of Sciences* 81 (1984): 801.

36. R.A. Muller, "Evidence for a Solar Companion Star," in *The Search for Extraterrestrial Life: Recent Developments, IAU Symposium*, 112, ed. M.D. Papagiannis (Dordrecht: D. Reidel, 1985), p. 233.

37. Robert A. Rohde and Richard A. Muller, "Cycles in Fossil Diversity," *Nature* 434 (2005): 209.

38. Mikhail V. Medvedev and Adrian L. Melott, "Do Extragalactic

Cosmic Rays Induce Cycles in Fossil Diversity?" *Astrophysical Journal* 664 (2007): 879.

39. Bertram Schwarzschild, "Varying Cosmic-Ray Flux May Explain Cycles of Biodiversity," *Physics Today* 60 (October 2007): 18.

40. Charles Darwin, *The Origin of Species and the Voyage of the Beagle* (New York: Alfred A. Knopf, 2003), p. 786.

CHAPTER 7. KEEP WATCHING THE SKIES!

1. "H. G. Wells in Woking," http://www.windowonwoking.org.uk/ sites/heritagewalks/Wells_Walk, based on work by Iain Wakeford (accessed February 12, 2008).

2. Horsell Common Preservation Society, http://www.horsell common.co.uk/sandpithc.htm (accessed October 18, 2006).

3. "Uncontacted Indian Tribe Found in Brazil's Amazon," *International Herald Tribune*, Associated Press, June 1, 2007, http://www.iht .com/articles/ap/2007/06/01/america/LA-GEN-Brazil-Indians.php (accessed December 4, 2007).

4. Carl Sagan, "Is Earth-Life Relevant? A Rebuttal," Planetary Society's Bioastronomy News, http://www.planetary.org/explore/ topics/search_for_life/seti/sagan2.html (accessed December 4, 2007).

5. Paul Davies, *Are We Alone?* (New York: Basic Books, 1995), p. 2.

6. Robert M. Torrance, ed., *Encompassing Nature: A Sourcebook* (Washington, DC: Counterpoint, 1998), p. 405.

7. Ibid., p. 724.

8. Nathaniel E. Green, "To the Editor of the Astronomical Register: Mars," *Astronomical Register* 17 (1879): 295.

9. Isaac Asimov, *Asimov's Biographical Encyclopedia of Science and Technology* (New York: Doubleday & Company, 1982), p. 470.

10. C. E. Burton, "The 'Canals' of Mars," *Astronomical Register* 18 (1880): 116.

11. E. W. Maunder, "The Canals on Mars," *Observatory* 11 (1888): 346.

12. L. Brenner, "On the Canals of Mars," *Observatory* 21 (1898): 296.

13. Asimov, *Asimov's Biographical Encyclopedia of Science and Technology*, p. 571.

14. William H. Pickering, "The Astronomy of Mars," *Popular Astronomy* 17 (1909): 497.

15. R. A. Proctor, *The Universe of Suns* (London: Chatto & Windus, Piccadilly, 1884), p. 163.

16. Percival Lowell, *Mars* (Boston: Houghton, Mifflin & Company, 1895), pp. 145–47.

17. Asimov, *Asimov's Biographical Encyclopedia of Science and Technology*, p. 556.

18. Robert Barker, "The 1937 Opposition of Mars," *Popular Astronomy* 46 (1938): 260.

19. The following paragraphs contain quotes and reports taken from the *Trenton Evening Times*, "Hoax Spreads Terror Here; Some Pack Up," pp. 1 and 2, October 31, 1938. These pages may be found at this fascinating Web site devoted to *The War of the Worlds*: http://www.war-ofthe-worlds.co.uk/index.html (accessed February 11, 2008).

20. Hadley Cantril, *The Invasion from Mars: A Study in the Psychology of Panic* (Edison, NJ: Transaction Publishers, 2005), pp. 67–68.

21. Dean Jamison, "Some Speculations on the Martian Canals," *Publications of the Astronomical Society of the Pacific* 77 (1965): 394.

22. Carl Sagan, "The Mariner IV Mission to Mars," *Astronomical Society of the Pacific Leaflets* 9 (1966): 366.

23. Harold Masursky et al., "Mars as Viewed from Mariner 9," *Icarus* 17 (1972): 289.

24. Robert G. Aitken, "Why Popular Interest in Mars?" *Astronomical Society of the Pacific Leaflets* 1 (1925): 6.

25. Carl Sagan, "Is Earth-Life Relevant? A Rebuttal," *Planetary Society's Bioastronomy News*, http://www.planetary.org/explore/topics/search_for_life/seti/sagan2.html (accessed December 4, 2007).

26. Giuseppe Cocconi and Philip Morrison, "Searching for Interstellar Communications," *Nature* 184 (1959): 844.

27. Frank D. Drake, "A Reminiscence of Project Ozma," http://www.bigear.org/vol1no1/ozma.htm (accessed December 6, 2007).

28. Robert H. Gray, "A VLA Search for the Ohio State 'Wow,'" *Astrophysical Journal* 546 (2001): 1177.

29. Frank Drake, "The E.T. Equation Recalculated," *Wired*, December 2004, http://www.wired.com/wired/archive/12.12/life.html (accessed November 6, 2007).

30. Marko Horvat, "Calculating the Probability of Detecting Radio Signals from Alien Civilizations," *International Journal of Astrobiology* 5 (2006): 143.

31. Marko Horvat, e-mail to author, November 8, 2007.

32. Ibid.

33. Geoffrey Bower, "Allen Telescope Array," *American Astronomical Society Meeting* 211 (2007): 210.66.03B.

34. Dennis Overbye, "Stretching the Search for Signs of Life," *New York Times*, October 11, 2007.

35. Abraham Loeb and Matias Zaldarriaga, "Eavesdropping on Radio Broadcasts from Galactic Civilizations with Upcoming Observatories for Redshifted 21 cm Radiation," *Journal of Cosmology and Astroparticle Physics* 01 (2007): 20.

36. Abraham Loeb, e-mail to author, November 8, 2007.

37. David Grinspoon, "Who Speaks for Earth?" *Seed* (November/December 2007): 70.

38. Editorial, "Ambassador for Earth," *Nature* 443 (2006): 606.

39. Seth Shostak, telephone interview with author, December 7, 2007.

40. N. S. Kardashev, "Transmission of Information by Extraterrestrial Civilizations," *Soviet Astronomy* 8 (1964): 217.

41. Robert Sanders, "Radio Telescope Array Dedicated to Astronomy, SETI," http://www.berkeley.edu/news/media/releases/2007/10/11_ata.shtml (accessed November 10, 2007).

42. Richard Dawkins, *The God Hypothesis* (New York: Houghton Mifflin Company, 2006), p. 202.

43. Ibid., p. 72.

44. Shostak, telephone interview, December 7, 2007.

45. OSETI I Workshop–13 (1993), http://www.coseti.org/1867 wk13.htm (accessed December 8, 2007).

46. Ibid.

47. Davies, *Are We Alone?* p. 50.

48. Jonathan D. Sarfati, "Bible Leaves No Room for Extraterrestrial Life," beliefnet (2006): http://www.beliefnet.com/story/140/story _14073.html (accessed December 9, 2007).

49. Jill Tarter, e-mail to author, November 20, 2007.

50. John Billingham, "Summary of Results of the Seminar on the Cultural Impact of Extraterrestrial Contact," *Publications of the Astronomical Society of the Pacific* 213 (2000): 670, 668.

51. Ibid., p. 672.

52. Tarter, e-mail to author, November 20, 2007.

53. You can keep up with the exo-planet count at the Extrasolar Planet Encyclopedia, http://exoplanet.eu/catalog.php (accessed December 13, 2007).

CHAPTER 8. EXOTICA

1. E. E. Barnard, "Observations of Halley's Comet at the Time of Its Nearest Approach to the Earth," *Astronomische Nachrichten* 185 (1910): 231.

2. J. Evershed, "Observations of the Tail of Halley's Comet Before and After the Day of Transit," *Monthly Notices of the Royal Astronomical Society* 70 (1910): 611.

3. Maxwell Hall, "Observations of Halley's Comet in Jamaica," *Monthly Notices of the Royal Astronomical Society* 70 (1910): 616.

4. Jeff Kanipe, *Chasing Hubble's Shadows: The Search for Galaxies at the Edge of Time* (New York: Hill and Wang, 2006), p. 107.

5. "Pioneer Anomaly," Planetary Society, http://www.planetary

.org/programs/projects/pioneer_anomaly/ (accessed February 19, 2008).

6. Priscilla Frisch and Jonathan D. Slavin, "Short-Term Variations in the Galactic Environment of the Sun," in *Solar Journey: The Significance of Our Galactic Environment for the Heliosphere and Earth*, ed. Priscilla C. Frisch (Dordrecht: Springer, 2006), p. 170.

7. Priscilla Frisch, "The Galactic Environment of the Sun," *American Scientist* 88 (2000): 52.

8. Research Consortium on Nearby Stars, the Nearby Stars Database, http://www.chara.gsu.edu/RECONS/ (accessed January 6, 2008).

9. Joan García-Sánchez et al., "Stellar Encounters with the Oort Cloud Based on *Hipparcos* Data," *Astronomical Journal* 117 (1999): 1042.

10. J.J. Matese, P.G. Whitman, and D.P. Whitmire, "Cometary Evidence of a Massive Body in the Outer Oort Cloud," *Icarus* 141 (1999): 354.

11. Douglas Spolyar, Katherine Freese, and Paolo Gondolo, "Dark Matter and the First Stars: A New Phase of Stellar Evolution," *Physical Review Letters* 100 (2008): 051101.

12. F.D. Kahn and L. Woltjer, "Intergalactic Matter and the Galaxy," *Astrophysical Journal* 130 (1959): 705.

13. T.J. Cox and Abraham Loeb, "The Collision between the Milky Way and Andromeda," arXiv0705.1170C (2007): 1.

14. "Massive Gas Cloud Speeding toward Collision with Milky Way," press release, National Radio Astronomy Observatory (January 11, 2008), http://www.nrao.edu/pr/2008/smithscloud/ (accessed January 15, 2008).

15. Gurtina Besla et al., "Are the Magellanic Clouds on Their First Passage about the Milky Way?" *Astrophysical Journal* 668 (2007): 949.

16. J.E. McClintock et al., "A Black Hole Greater than 6 Solar Masses in the X-Ray Novae XTE J1118+480," *Astrophysical Journal* 551 (2000): L147.

17. Jeffrey E. McClintock and Ronald Remillard, "Black Hole Binaries," in *Compact Stellar X-Ray Sources*, ed. W.H.G. Lewin and M. van der Klis (Cambridge: Cambridge University Press, 2003), p. 157.

18. I. F. Mirabel et al., "A High-Velocity Black Hole on a Galactic-Halo Orbit in the Solar Neighborhood," *Nature* 413 (2001): 140.

19. Ibid., p. 141.

20. Govert Schilling, "Rogue Black Holes May Abound in Milky Way," *Science*Now Daily News, January 9, 2008, http://sciencenow .sciencemag.org/cgi/content/full/2008;109/4 (accessed January 16, 2008).

21. P. Hut and M. J. Rees, "How Stable Is Our Vacuum?" *Nature* 302 (1983): 508.

22. Sidney Coleman and Frank De Luccia, "Gravitational Effects On and Of Vacuum Decay," *Physical Review D* 21 (1980): 3305.

23. R. L. Jaffe et al., "Review of Speculative 'Disaster Scenarios' at RHIC," *Reviews of Modern Physics* 72 (2000): 1125.

24. Renata Kallosh et al., "Observational Bounds on Cosmic Doomsday," *Journal of Cosmology and Astroparticle Physics* 10 (2003): 15.

25. Max Tegmark and Nick Bostrom, "Is a Doomsday Catastrophe Likely?" *Nature* 438 (2005): 754.

26. Robert R. Caldwell, Marc Kamionkowski, and Nevin N. Weinberg, "Phantom Energy: Dark Energy with w<-1 Causes a Cosmic Doomsday," *Physical Review Letters* 91 (2003): 071301.

CHAPTER 9. THE ULTIMATE COSMIC CONNECTION

1. Brandon Carter, "Large Number Coincidences and the Anthropic Principle in Cosmology," in *Confrontation of Cosmological Theories with Observational Data*, ed. M. S. Longair (Boston: Reidel, 1974), p. 132.

2. Alex Vilenkin, *Many Worlds in One: The Search for Other Universes*, (New York: Hill and Wang, 2006), p. 133.

3. Geoff Brumfiel, "Our Universe: Outrageous Fortune," *Nature* 439 (2006): 12.

4. Richard Swinburn, "Argument from the Fine-Tuning of the

Universe," in *Physical Cosmology and Philosophy*, ed. John Leslie (New York: Macmillan, 1990), p. 172.

5. Lee Smolin, "Scientific Alternatives to the Anthropic Principle," in *Universe or Multiverse?* ed. Bernard Carr (Cambridge: Cambridge University Press, 2007), p. 323.

6. Heinz R. Pagels, "A Cozy Cosmology," *Sciences* 25 (1985): 35.

7. Smolin, "Scientific Alternatives to the Anthropic Principle," p. 37.

8. Stephen J. Gould, "Mind and Supermind," in *Physical Cosmology and Philosophy*, ed. John Leslie (New York: Macmillan, 1990), p. 183.

9. N. Gorlova et al., "Debris Disks in NGC 2547," *Astrophysical Journal* 670 (2007): 533.

10. F. Hammer et al., "The Milky Way Is an Exceptionally Quiet Galaxy: Implications for the Formation of Spiral Galaxies," *Astrophysical Journal* 662 (2007): 322.

11. M. E. Beer at al., "How Special Is the Solar System?" *Monthly Notices of the Royal Astronomical Society* 354 (2004): 763.

12. D. Malmberg et al., "Is Our Sun a Singleton?" in the proceedings of IAU Symposium 246, *Dynamical Evolution of Dense Stellar Systems*, ed. E. Vesperini (2008): 119.

13. B. S. Gaudi et al., "Discovery of a Jupiter/Saturn Analog with Gravitational Microlensing," *Science* 319 (2008): 927.

14. J. K. Webb et al., "Further Evidence for Cosmological Evolution of the Fine Structure Constant," *Physical Review Letters* 87 (2001): 091301.

INDEX

accelerating universe, 244
Adhemar, Joseph, 19, 28
Agassiz, Louis, 17–18, 28
Aiguille des Grands Charmoz, 39
Aitken, Robert G., 197
Alpha Centauri, 163, 229
Andromeda Galaxy, 232–33
anomalous X-ray pulsars (AXPs),
    143
anthropic principle, 254, 256
    arguments against, 256–58
Apollo missions, 65–66
Apophis. *See* asteroids
Arcturus, 130
Arve River, 39, 44
asteroids
    Apophis, 98
    asteroid belt, 98
    Baptistina asteroid family, 115
    compared with meteoroids, 96

Gaspra, 97
Ida, 97
impact scenario, 120–24
mitigation of, 111–13
Near Earth Asteroids (NEAs),
    107
Near Earth Comets (NECs),
    107
Near Earth Objects (NEOs),
    106
near misses, 99
Potentially Hazardous Objects
    (PHOs), 107
probability of mortality, 107–
    108
probability of Earth impact,
    96–99
search for, methodologies, 109–
    11
astrometry, 130

Barker, Robert, 195
Barnard, Edward Emerson, 155, 223
Barnard's Star, 163
Beta Scorpii, 155
Billingham, John, 209
black holes
    microquasars, 240–42
    rogue, 242
Bonanay, 44
Brenner, Leo, 193–94
brown dwarfs, 227, 230
Brückner, Eduard, 22
Bruegel, Pieter the Elder, 47
Bruegel, Pieter the Younger, 47
Burton, C. E., 193

Carrington, Richard, 75
Carter, Brandon, 255
catastrophism, 17
Chamonix, 39
Chapman, Clark, 107–108
Chicxulub impact, 115
Clarke, Arthur C., 125, 209
Cocconi, Giuseppe, 198
Comet Halley, 224
Comet LINEAR, 12
comets, 11–12, 103, 113–14, 179, 224, 225, 229–30
Comet Shoemaker-Levy, 114
Contact (motion picture), 218
cosmic coincidences, 252–54
    changes over time, 261–63

cosmic rays, 55
    climate, 171, 175
    role in mass extinctions, 132, 181–82
    solar activity, 56
Cox, Thomas, 235–36
Croll, James, 19, 28
Cuvier, Georges, 177

dark energy, 245, 248
Darwin, Charles, 20, 183
Davidson, Kris, 135–36, 138, 139–40
Davies, Paul, 216
Dawkins, Richard, 213
Department of Homeland Security, 81–82
Dickens, Charles, 47
Double Cluster, 161
Douglas, Andrew Ellicott, 194
Draco Molecular Cloud, 169
Drake, Frank, 198
Drake equation, 201–202, 212, 219
drumlins, 15–16
Duncan, Robert, 144–48

Eagle Nebula, 161
Earth, magnetosphere, 171, 229
eccentricity of Earth's orbit, 30–35
Eddy, Jack A., 54–56
Ehman, Jerry, 199–200
Epsilon Eridani. See sunlike stars

ergs, 70

Erwin, Doug, 131

Eta Carinae
axial orientation in Milky Way, 138
as an "extreme object," 136
mass loss, 136–37
physical properties, 134, 137
variations, 134–35

E.T.: The Extra-Terrestrial (motion picture), 207

Evershed, John, 224

extrasolar planets, 210, 220

Finsteraahorn, 18

Flammarion, Nicolas, 194

flares. *See* solar activity

Forbes bands, 40

fossils, 35, 177–78

Frankenstein (Shelley), 40

Fruchter, Andrew, 141

Gaia (astrometry mission), 133

Galileo, 53

Gamma-ray Burst Optical Studies with HST (GOSH), 141

gamma-ray bursts (long- and short-duration), 140–43
detected by spacecraft, 144
mutagenetic effects, 183

Geostationary Operational Satellites, 74–75

Giant Molecular Clouds, 159, 176, 182

glaciation, 15–16

glaciation periods, 22–23

glaciers
Aar, 18
Argentière, 18
Jura Mountains, 17, 28
Les Bossons, 40
Mer de Glace (Glacier des Bois), 40–46
exorcism, 45
rapid advancement, 43–44
retreat, 46
Wisconsin, 15

global warming, 42, 58–60

globular clusters, 142, 161, 241

Gliese 710 (star), 230

Gould, Stephen J., 257–58

Gould's Belt, 130–31

Gum Nebula, 165–66

Halley, Edmund, 129

Hartmann, Johannes, 154

Hawks Nest, New South Wales, 157

heavy water, 12

Heger, Mary Lea, 155

Helix Nebula, 166

Hipparchus (astronomer), 129

Hipparcos (astrometric satellite), 61, 130, 133, 229

Hipparcos Catalogue, 164

Hiroshima bomb, yield of, 70–71
Hodgson, R., 75
Holley-Bockelmann, Kelly,
    242–43
Homunculus (Eta Carinae),
    137–38
Hooke, Robert, 177
Horsell Common, United
    Kingdom, 186
Horvat, Marko, 202–204
Hubble Space Telescope, 92
hypernovae, 136

ice ages, 18–20
    Sun's motion through Galaxy
        as cause for, 158–59,
        173–74, 176–77, 179
ice core evidence (various), 49,
    55, 76, 117, 148–49, 173, 176
impact craters
number on Earth, 115–16
Indian Ocean tsunami, 108
International Quaternary Associ-
    ation (INQUA), 20
interstellar clouds
    Apex Cloud, 228–29
    cloudlets, 227, 229
    encounters with Sun, 158, 163,
        228–30
    G Cloud, 228–29
    high-velocity clouds, 169
    in general, 155

Kardashev, Nikolai, 212
Kennett, James P., 117–18
Kepler, Johannes, 52
Kepler's Star, 149
Koch, Howard, 188, 195

Lagoon Nebula, 161
Lagrange, Joseph Louis, 19
Large Magellanic Cloud, 148,
    157, 233, 239
Late Ordovician period, 147
Le Bois, 44
Le Châtelard, 44
Les Drus, 43
Les Rousier, 44
Le Verrier, Urbain Jean Joseph, 19
Lightest Supersymmetric Particle
    (LSP), 231
Little Ice Age, 46–49
    effects of, 47–48
    as neoglaciation, 46
Local Bubble, 130, 132, 166, 168
Local Interstellar Cloud, 159, 168
local standard of rest, 163
Loeb, Abraham, 207, 209, 235–36
Lowell, Percival, 194
Low Frequency Array (LOFAR),
    206
Lyell, Sir Charles, 17, 20
Lyman-alpha radiation, 153–55

Magellanic Clouds, 142, 239
Magellanic Stream, 239

magnetars, 143–48
Mars
  glaciers, 36
  life on Mars, 191
  solar radiation on, 66
Mars Global Surveyor, 36
Martians, 185, 187, 194
Mariner missions, 196
Marusek, James A., 120
mass extinctions, 177–83
  Darwin's views on, 183–84
Matterhorn, 17
Matthes, François, 46
Maunder, Edward Walter, 53–54
  "canals" on Mars, 193
Maunder minimum, 53–54
Medieval Warm Epoch, 56–57
Medvedev, Mikhail V., 180–82
Melott, Adrian L., 180–82
Meteor Crater, Arizona, 96
  as Barringer crater, 105–106
meteorites
  alleged deaths and injuries, 91–
    92
  damage to spacecraft, 92–93
  probabilities of death or injury
    by, 95–96
meteoroids, 96
Milankovitch, Milutin
  astronomical cycles, 27, 28–35
  childhood and education, 25–26
  confrontation with Penck, 22–23
  flaws in theory, 36–37

mathematical theory of cli-
    mate, 26
  theory of ice ages, 20–22
Mileura Wide-Field Array
    (MWA), 206
Milky Way
  collision with Andromeda
    Galaxy, 232–35
    effect on the Milky Way, 236–
      37
    effect on the Sun, 235–36
  collision with Smith's Cloud,
    237–38
  passage through intergalactic
    space, 180–82
  as seen from Earth, 156–58
  spiral arms, 176
  Sun's passage through, 158
Mintaka, 155
Mont Blanc, 39
Montenvers, 40
Moon, 258–59
  change in orbit, 261–62
Morrison, David, 107
Morrison, Philip, 198
Muller, Richard A., 179–80
Murchison, Sir Roderick, 178

Near Earth Asteroids. *See* asteroids
Near Earth Comets. *See* asteroids
Near Earth Objects. *See* asteroids
Nemesis, 179, 230

obliquity of Earth's axis, 28–30

Odessa Meteor Crater, 107

Oerlemans, Johannes, 48–49

Oort Cloud, 179, 230, 242, 263

Orion (constellation), 131

Orion Spur, 165

Ozma. *See* Project Ozma

Pagels, Heinz, R., 256–57

PAHNs, 11

Peekskill meteor event, 87–90
  meteoroid's orbit and trajectory,
  103

Penck, Albrecht, 23–25

Permo-Triassic extinction, 148,
  178, 182

Pickering, William H., 194

Pioneer anomaly, 227–28

planetary nebulae, 166

Potentially Hazardous Objects.
  See asteroids

precession, 31–34

Proctor, Richard Anthony, 194

Procyon, 130

Project Ozma, 198–99

Raup, David M., 179

Ray, John, 177

Redfield, Seth, 133

Rees, Sir Martin, 226

Reuven Ramaty High Energy
  Solar Spectroscopic Imager
  (RHESSI), 70

Roe, Gerard, 37–38

Rubenstein, Eric, 79

Sagan, Carl, 189, 197

Schaefer, Bradley, 79

Schiaparelli, Giovanni, 192

Schindewolf, Otto, 131

Scorpius-Centaurus association,
  131–33, 168–69

Search for Extraterrestrial Intelligence (SETI), 201
  Allen Array, 204
  limits to radio communication
    by advanced civilizations,
    202–204
  "piggyback" approach to SETI,
    204

Search for Extraterrestrial Radio
  Emissions from Nearby
  Developed Intelligent Popu-
  lations (SERENDIP), 205

societal effects of contact with
  alien intelligence, 213–19

types of civilizations (Kardashev), 212

Sepkoski, John, Jr., 179

Shapley, Harlow, 173

Shostak, Seth, 210, 214–15

Sikhote-Alin fall, 102

Sirius, 135

Small Magellanic Cloud, 157, 239

soft gamma-ray repeaters (SGRs),
  143

Solanki, S. K., 60
solar activity, 54–56, 60, 68
  coronal mass ejections (CME),
    68–71, 73–74, 76, 80, 83
  flares (in general), 68
    danger to astronauts, 65–66
    1859 flare, 75–77
    energy yield, 70–71
    exceptional flare events, 67,
      75–77
    formation of, and relation to,
      coronal mass ejections,
      70–73
    flares on sunlike stars, 78–79
    super flares, 77
    threat to major power grids,
      80–82
    X-ray flares, 68, 70–71, 74
solar antapex, 163
solar apex, 162
solar constant, 50
solar heliosphere, 159, 166–71
Solar Heliospheric Observatory
    (SOHO), 71, 74
solar magnetosphere, 68, 77
Solar Radiation and Climate
    Experiment (SORCE), 70
space debris, 93–94
  amount generated by destruc-
    tion of Fengyun-1C, 95
Spörer, Gustav, 53
Spörer minimum, 53–54
Square Kilometer Array, 206

stars
  with companions, 79–80
  dark stars, 231
  first stars in the universe, 206,
    231
  magnetars. *See* magnetars
  massive stars, 172
  metals, 141–42
  motions through space, 128, 130
  near the Sun, 229
  neutron stars, 142, 241
  pulsating, 51
  as seeds of life, 11
  supernovae. *See* supernovae
  variable stars, 50
Star Trek (TV series), 127, 164,
    202, 219
Stradivari, Antonio, 49
Struiver, Minze, 55
Sun
  death of, 83–84
  faculae, 52
  magnetic field, 55
  sunspots, 51
    sunspot records, 52–54
  as a variable star, 51–52, 58–60
sunlike stars, 60–62, 78
  Epsilon Eridani, 198–99
  18 Scorpii, 61
  HD 98618, 61
  RS Canum Venaticorum stars,
    78
  Tau Ceti, 198

supernovae, 125–27
  danger zone, 129
  mutagenetic effects, 183
  nearby supernova candidates,
    129, 134
  rate of, 127
  recent in Milky Way, 148
  supernova remnants, 126
  SN 2006gy, 150
  Type Ia, 244
  Vela supernova, 166
  Vela Jr., 149
Svensmark, Henrik, 173, 175–76

Tarter, Jill, 218, 220
Tatel Telescope, Green Bank,
  West Virgina, 198
Tau Ceti. *See* sunlike stars
Taurid Complex, 114
Tempel, Gugliemo, 69
Thames River, 47
tree rings, 55–56, 60
Trifid Nebula, 161
Tsar Bomba, nuclear bomb, 121
Tsurutani, Bruce, 77
Tunguska event, 100–101
  exotic explanations, 105
  frequency of, 105
  origin of, 114
  search for impactor, 102, 104
  whether asteroid or comet,
    101–102
Tycho's Star, 149

uniformitarianism, 17
US Space Surveillance Network,
  93

vacuum
  cause of collapse, 246–47
  collapse of, 243, 245–46
  vacuum energy, 245, 256
variable stars. *See* stars
Virgo Cluster of galaxies, 180

War of the Worlds, The (novel), 185,
  191
War of the Worlds, The (radio
  broadcast), 195–96
  broadcasts in Quito, Ecuador,
    and Santiago, Chile, 196
Wegener, Alfred, 27, 28
Welles, Orson, 188, 195
Wells, H. G., 185, 186, 187
Wolf, Max, 155
Wolf minimum, 54
Wow! signal, 199–201

Yarkovsky effect, 113
Younger Dryas, 117–19

Zaitsev, Alexander, 209